Studies in Big Data

Volume 174

Series Editor
Janusz Kacprzyk, Polish Academy of Sciences, Warsaw, Poland

The series "Studies in Big Data" (SBD) publishes new developments and advances in the various areas of Big Data- quickly and with a high quality. The intent is to cover the theory, research, development, and applications of Big Data, as embedded in the fields of engineering, computer science, physics, economics and life sciences. The books of the series refer to the analysis and understanding of large, complex, and/or distributed data sets generated from recent digital sources coming from sensors or other physical instruments as well as simulations, crowd sourcing, social networks or other internet transactions, such as emails or video click streams and other. The series contains monographs, lecture notes and edited volumes in Big Data spanning the areas of computational intelligence including neural networks, evolutionary computation, soft computing, fuzzy systems, as well as artificial intelligence, data mining, modern statistics and Operations research, as well as self-organizing systems. Of particular value to both the contributors and the readership are the short publication timeframe and the world-wide distribution, which enable both wide and rapid dissemination of research output.

The books of this series are reviewed in a single blind peer review process.

Indexed by SCOPUS, EI Compendex, SCIMAGO and zbMATH.

All books published in the series are submitted for consideration in Web of Science.

Francisco Chinesta · Elías Cueto ·
Victor Champaney · Chady Ghnatios ·
Amine Ammar · Nicolas Hascoët ·
David González · Icíar Alfaro ·
Daniele Di Lorenzo · Angelo Pasquale ·
Dominique Baillargeat

A Gentle Introduction to Data, Learning, and Model Order Reduction

Techniques and Twinning Methodologies

Authors
See next page

ISSN 2197-6503 ISSN 2197-6511 (electronic)
Studies in Big Data
ISBN 978-3-031-87571-7 ISBN 978-3-031-87572-4 (eBook)
https://doi.org/10.1007/978-3-031-87572-4

This work was supported by Arts et Métiers Institute of Technology and University of Zaragoza.

© The Editor(s) (if applicable) and The Author(s) 2025. This book is an open access publication.

Open Access This book is licensed under the terms of the Creative Commons Attribution-NonCommercial-NoDerivatives 4.0 International License (http://creativecommons.org/licenses/by-nc-nd/4.0/), which permits any noncommercial use, sharing, distribution and reproduction in any medium or format, as long as you give appropriate credit to the original author(s) and the source, provide a link to the Creative Commons license and indicate if you modified the licensed material. You do not have permission under this license to share adapted material derived from this book or parts of it.
The images or other third party material in this book are included in the book's Creative Commons license, unless indicated otherwise in a credit line to the material. If material is not included in the book's Creative Commons license and your intended use is not permitted by statutory regulation or exceeds the permitted use, you will need to obtain permission directly from the copyright holder.
This work is subject to copyright. All commercial rights are reserved by the author(s), whether the whole or part of the material is concerned, specifically the rights of translation, reprinting, reuse of illustrations, recitation, broadcasting, reproduction on microfilms or in any other physical way, and transmission or information storage and retrieval, electronic adaptation, computer software, or by similar or dissimilar methodology now known or hereafter developed. Regarding these commercial rights a non-exclusive license has been granted to the publisher.
The use of general descriptive names, registered names, trademarks, service marks, etc. in this publication does not imply, even in the absence of a specific statement, that such names are exempt from the relevant protective laws and regulations and therefore free for general use.
The publisher, the authors and the editors are safe to assume that the advice and information in this book are believed to be true and accurate at the date of publication. Neither the publisher nor the authors or the editors give a warranty, expressed or implied, with respect to the material contained herein or for any errors or omissions that may have been made. The publisher remains neutral with regard to jurisdictional claims in published maps and institutional affiliations.

This Springer imprint is published by the registered company Springer Nature Switzerland AG
The registered company address is: Gewerbestrasse 11, 6330 Cham, Switzerland

If disposing of this product, please recycle the paper.

Francisco Chinesta
Arts et Metiers Institute of Technology
Paris, France

Victor Champaney
Duoverse
Paris, France

Amine Ammar
Arts et Métiers Institute of Technology
Angers, France

David González
Aragon Institute for Engineering Research
Universidad de Zaragoza
Zaragoza, Spain

Daniele Di Lorenzo
Duoverse
Paris, France

Dominique Baillargeat
University of Limoges
Limoges, France

Elías Cueto
Aragon Institute for Engineering Research
Universidad de Zaragoza
Zaragoza, Spain

Chady Ghnatios
University of North Florida
Jacksonville, FL, USA

Nicolas Hascoët
Arts et Metiers Institute of Technology
Paris, France

Icíar Alfaro
Aragon Institute for Engineering Research
Universidad de Zaragoza
Zaragoza, Spain

Angelo Pasquale
Duoverse
Paris, France

Acknowledgments This research is part of the DesCartes programme and is supported by the National Research Foundation, Prime Minister Office, Singapore, under its Campus for Research Excellence and Technological Enterprise (CREATE) programme.

The authors also acknowledge the support from the Keysight Technologies, RTE and SKF chairs at ENSAM Institute of Technology.

This work was supported by the Spanish Ministry of Science and Innovation, AEI/ 10.13039/501100011033, through Grant number PID2023-147373OB-I00, and by the Ministry for Digital Transformation and the Civil Service, through the ENIA 2022 Chairs for the creation of university-industry chairs in AI, through Grant TSI-100930-2023-1.

This project has received funding from the European Union Horizon 2020 research and innovation programme under the Marie Skłodowska-Curie grant agreement No. 956401 (XS-Meta).

Declarations

- The authors have no competing interests to declare that are relevant to the content of this manuscript.
- All authors declare they have contributed equally to this work and there is no interest conflict.
- Conflict of interest: Authors have no financial or non-financial interests to disclose.
- Competing interests: The authors have no competing interests to declare.
- Ethics approval: Not applicable.
- Consent for publication: All authors consent the publication of this article.
- Availability of data and materials: Not applicable.
- Code availability: Not applicable.

Contents

1	**Introduction**	1
2	**Extended Summary**	3
	2.1 From Virtual to Digital Twins	5
	2.2 In Between Models and Data	7
	2.2.1 Physics-Informed Learning	8
	2.2.2 Physics-Augmented Learning	8
	2.2.3 Discussion	8
	2.2.4 The Global Picture	9
	2.3 Physics in Real-Time: Model Order Reduction	10
	2.4 Data and Learning	10
	2.5 Hybridation	12
	2.5.1 Solution Enrichment	12
	2.5.2 Model Enrichment	12

Part I Around Data

3	**Intrinsic Dimensionality of a Data Set and Manifold Learning**	17
	3.1 Principal Component Analysis, PCA, and Its Local Counterpart, ℓPCA	18
	3.2 Multidimensional Scaling, MDS	19
	3.3 Kernel Principal Component Analysis, kPCA	19
	3.4 Locally Linear Embedding, LLE	20
	3.5 t-Distributed Stochastic Neighbor Embedding, tSNE	21
	3.5.1 Stochastic Neighbor Embedding, SNE	21
	3.5.2 Coming Back to the tSNE	22
	3.6 PCA on Qualitative Variables	22
	3.7 PCA in Presence of Missing Data	23
	References	23

4	**Intrinsic Dimensionality and Autoencoders**	25
	4.1 Sparse Autoencoders	26
	4.2 Variants	27
	4.2.1 Variational Autoencoders	27
	4.2.2 Denoising Autoencoders	27
	4.2.3 Contractive Autoencoders	28
	References	28
5	**Tensor Formats and Tensor Decompositions**	29
	5.1 SVD and HOSVD Versus PGD	29
	5.2 CUR Decomposition	30
	5.3 Cross Approximation	30
	References	31
6	**Dictionary Learning**	33
	Reference	34
7	**Time Series: Metrics and Alignment**	35
	7.1 Standard Metrics for Time Series	35
	7.2 Dynamic Time Warping	36
	7.3 Optimal Transport and Topological Data Analysis	37
	References	37
8	**Data Separation: Independent Component Analysis**	39
	References	40
9	**Topological Data Analysis**	41
	9.1 Illustrating TDA on a Time Series	41
	9.2 Filtration and Homology Group	42
	9.2.1 Rips Filtration	43
	9.2.2 Other Filtrations	44
	References	46
10	**Compressed Sensing**	47
	10.1 Sparsity	47
	10.2 Sparse Sensing	47
	References	48
11	**Some Issues in Data Interpolation**	51
	11.1 Complex-Valued Variables	51
	11.1.1 A Simple Phase Unwrapping Procedure	52
	11.2 Interpolating Reduced Bases: The Grassmann Manifold	52
	References	54
12	**Random Variables: Probability, Statistics and Bayesian Learning**	55
	12.1 Probabilities	55
	12.1.1 Discrete Probability Distributions	55
	12.1.2 Continuous Probability Distributions	56

		12.1.3 Moments	56
		12.1.4 Other General Properties	56
	12.2	Random Vectors	56
	12.3	Statistical Inference	57
		12.3.1 Point Estimate	57
		12.3.2 Interval Estimate	57
		12.3.3 Statistical Tests	58
	12.4	Bayesian Learning	58
		12.4.1 Naive Bayes	58
	References		59
13	**Random Variables: Polynomial Chaos, PC**		61
	13.1	Functions of Random Variables	61
		13.1.1 The Univariate Case	61
		13.1.2 Polynomial Chaos, PC	61
		13.1.3 The Multivariate Case	62
	13.2	Polynomial Chaos Expansion for Random Fields	62
	13.3	Uncertainty Propagation, UP	63
		13.3.1 Intrusive Procedure	63
		13.3.2 Non-intrusive Procedure Based on PC Expansion	64
		13.3.3 Non-intrusive Procedure Based on Regression	64
	References		65
14	**Radom Variables: Gaussian Processes, GP**		67
	14.1	Multivariate Gaussian Distribution	67
	14.2	Gaussian Process	68
	References		68
15	**Analysis of Variance, ANOVA**		71
	15.1	ANOVA Decomposition	71
	15.2	Sensitivity Analysis: Sobol Coefficients	72
	15.3	Anchored ANOVA	72
	References		72

Part II   Around Learning

16	**Data Classification and Clustering**		77
	16.1	Unsupervised Clustering: k-Means	77
	16.2	Supervised Data Classification	77
		16.2.1 Support Vector Machines, SVM	77
		16.2.2 Decision Trees and Random Forest	78
	References		79
17	**Boosting Algorithms**		81
	17.1	AdaBoost	81
	17.2	Gradient Boosting and Its Stochastic Counterpart	82
	References		83

18	**Learning Modalities**		85
	18.1	Self-supervised Learning	85
	18.2	Semi-supervised Learning	86
	18.3	Transfer Learning	87
	18.4	Reinforcement Learning, RL	88
	References		90
19	**Regression: Basics**		91
	19.1	Polynomial Regression	91
	19.2	Kriging	92
	19.3	Support Vector Regression, SVR	93
	19.4	Likelihood-Based Regressions for Noisy Data	94
	References		96
20	**Neural Network Based Machine Learning Techniques**		97
	20.1	From the Neuron to Deep Neural Networks	97
	20.2	The Universal Approximation Theorem	99
	20.3	Learning Operators: *DeepONets*	100
	20.4	Convolutional NN, CNN (ConvNet)	101
	20.5	Graph Neural Network, GNN, and Message Passing, MP	102
	20.6	Recurrent Neural Networks, rNN	104
	20.7	Long Short Time Memory, LSTM	105
	20.8	Gated Recurrent Unit, GRU	106
	20.9	Reservoir Computing	107
	20.10	Residual Nets	107
	20.11	NeuralODE	108
	20.12	Transformers	108
	20.13	Physics-Informed Neural Networks, PINN	110
	20.14	Thermodynamics Informed Neural Networks, TINN	110
	20.15	Generative Adversarial Network, GAN	113
	20.16	Nonlinear Autoregressive Network with Exogenous Inputs, NARX	114
	20.17	Multimodal Learning Based on Boltzmann Machines, BM	115
	References		118
21	**Other Machine Learning Techniques**		121
	21.1	Code2Vect	121
	21.2	Sparse Identification for Nonlinear Dynamical Systems, SINDy	122
	21.3	Sparse PGD Based Regressions, sPGD	123
	21.4	The Koopman Operator	124
	21.5	Dynamic Mode Decomposition, DMD	125
	21.6	Incremental DMD, iDMD	126
		21.6.1 Progressive Greedy Constructor, PGC	127
		21.6.2 Rank-p Constructor, RPC	127
		21.6.3 Reduced Formulation	128

		21.6.4	Noise Filtering	128
		21.6.5	Nonlinear Models	128
	References			128

Part III Around Reduction

22 From Discretization to Model Order Reduction 133
    22.1   Reduced Order Modelling 134
    22.2   More on PGD-Based Separated Representations 135
    Reference 136

23 Proper Orthogonal Decomposition and Reduced Basis 137
    23.1   Proper Orthogonal Decomposition Based Model Order
            Reduction 137
    23.2   Snapshot-POD 138
    23.3   Hyper-Reduction 138
    23.4   Compressed Sensing Based Adaptive Mode Selection 138
    23.5   Reduced Basis, RB 139
    23.6   Gappy-POD 139
    23.7   Nonlinear Models 140
            23.7.1   Nonlinearities Interpolated: The Discrete
                      Empirical Interpolation Method, DEIM 140
            23.7.2   Trajectory Piece-Wise Linear, TPWL 141
    23.8   Grassman and Barycentric Interpolation 143
            23.8.1   Grassman Manifold 143
            23.8.2   From Grassman to Barycentric Interpolation
                      of Reduced Bases 144
    References 145

24 The Proper Generalized Decomposition 147
    24.1   Extended Separated Representations 147
    24.2   Nonlinear Models 149
            24.2.1   Incremental Linearization 149
            24.2.2   Newton Linearization 150
            24.2.3   Enhancing Iteration Procedures 150
            24.2.4   Asymptotic Numerical Method, ANM 150
            24.2.5   Discrete Empirical Interpolation Method, DEIM 151
    References 152

25 Design of Experiments and Surrogate Models 155
    25.1   Design of Experiments, DoE 155
            25.1.1   The Multifactorial Approach of Plackett
                      and Burman and Factors Influence 155
            25.1.2   Fisher Information Matrix, FIM 156

		25.1.3 Quadratures and Latin-Hypercube	157

    25.2   Surrogate Models Based on Response Surface
Methodology, RSM 157
References 158

26 Parametric Models: POD Based Surrogates 159
    26.1   POD with Interpolation 159
    26.2   Extension to Multi-parametric Scenarios 160
    Reference 161

27 Parametric Models: PGD-Based Surrogates 163
    27.1   Sparse Subspace Learning 163
    27.2   Collocation-Based Sparse PGD and Its Regularized
Variants 165
            27.2.1 Regularized Formulations 167
            27.2.2 Smaller Than the L1 Norm 167
    27.3   Projection-Based sPGD 169
    27.4   From ANOVA-Based Metamodelling to ANOVA-Based
sPGD 171
            27.4.1 ANOVA-Based sPGD 171
            27.4.2 Statistics 172
    27.5   Multi-PGD, mPGD 172
            27.5.1 Intrusive Framework 172
            27.5.2 Non-intrusive Framework 173
    27.6   Parametric Transfer Function, PTF 174
    References 175

28 Quantities Of Interest 177
    28.1   Parametric QoI 177
    28.2   Uncertainty Quantification 178
    28.3   Adjoint Method 179
    Reference 180

29 Projection and Transport 181
    29.1   Riemannian Projectors 181
            29.1.1 Ricci Flow Based Conformal Mapping and Its
Associated Reduction Procedure 182
    29.2   Smart Mapping 182
    29.3   From Optimal Transport to Parametric Optimal Transport 184
    References 186

30 Multiscale 187
    30.1   Partition of Unity-Based Enrichment 187
    30.2   Direct Separated Decompositions 188
    References 189

31 Dynamics 191
- 31.1 Transient Dynamics in the Physical Space 191
- 31.2 Mass Lumping and Modal Analysis 192
- 31.3 Harmonic Analysis 192
- 31.4 Harmonic-Modal Hybrid Analysis 193
  - 31.4.1 Nonlinear Models 194
- References 195

32 Space Separation 197
- 32.1 In-Plane-Out-of-Plane Separated Representation of a Multilayered Plate 197
- 32.2 Addressing Non-separable Domains: Smart Mapping 198
- 32.3 NURBS-Based PGD 200
  - 32.3.1 NURBS-Based Geometry Description 200
- 32.4 Non-intrusive Space Separation 202
- References 203

Part IV Around Data Assimilation and Twining

33 Data Assimilation, Inverse Analysis and Control 207
- 33.1 Optimal Control 207
- 33.2 Assimilation by Tikhonov Regularization of PGD-Based Parametric Solutions 209
  - 33.2.1 Assimilation Based on the Complete PGD Solution 209
  - 33.2.2 Assimilation Based on a PGD-Based Reduced Basis 210
- 33.3 Optimal Sensor Placement for PGD-Based Data Assimilation 211
- 33.4 Bayesian Inverse Analysis and Data-Assimilation 211
  - 33.4.1 PGD-Likelihood Maximization in Absence of Priors 211
  - 33.4.2 Accounting for Priors in Bayesian Settings 212
- 33.5 Data-Assimilation Based on Kalman Filters 212
  - 33.5.1 Extended Kalman 214
- 33.6 PGD-Based Parametric Inverse Impulse Response, IIR 215
  - 33.6.1 PGD-Based Direct Impulse Response, DIR 215
  - 33.6.2 Real-Time Monitoring 216
  - 33.6.3 Inverse Impulse Response 216
  - 33.6.4 Data-Driven Regularization 217
- 33.7 Inverse Analysis Based on the Reciprocity Principle, RP 217
  - 33.7.1 Symmetric Operators: The Elastic Problem 217
  - 33.7.2 Non-symmetric Operators: The Heat Equation 218
- References 218

34	**The Twin Continuum**		221
	34.1	Fully Physics-Based Modelling	221
	34.2	Fully Data-Driven Modeling	222
	34.3	Just in Between Both Paradigms	223
		34.3.1 Physics-Aware Learning	223
		34.3.2 Physics-Augmented Learning	224
	References		225
35	**Conclusion**		227

Chapter 1
Introduction

Already in 2017, humanity produced in two years more data than in its entire history. By that time, large scientific infrastructures such as the CERN, in Switzerland, produced some 30 petaBytes per year. This book is a gentle introduction to the fascinating world of machine learning. We focus in detail on techniques belonging to the even more fascinating discipline of scientific machine learning, a subset of the former that tries to extract insight from scientific data.

This field of scientific machine learning emerges in the context of the so-called fourth paradigm of science.[1] Indeed, science has become increasingly dependent of massive amounts of data generated by large scientific infrastructures. All these data must be collected, curated and analyzed efficiently. This book is devoted to the description of the most important algorithms that allow us to do so.

On the other hand, the industry 4.0 paradigm, or fourth industrial revolution, is revolutionizing manufacturing by an increasing interconnectivity of every asset, artificial intelligence, and smart robotic automation. Scientific machine learning constitutes an indispensable ingredient of this industrial revolution. It allows, notably, for the development of digital twins, virtual copies of physical entities that receive continuously information from its physical counterpart and process it, analyzes, simulates future scenarios and make decisions based on rigorous information and state-of-the-art models of the physical processes taking place. Even more sophisticate, the so-called hybrid or cognitive digital twins are able to detect anomalies in the built-in models, if discrepancies with the data persist, and to correct themselves by developing more efficient and accurate data-based models.

---

[1] The fourth paradigm of science emerges as a concept after the book The Fourth Paradigm: Data-Intensive Scientific Discovery. Tony Hey, Stewart Tansley, Kristin Tolle, Jim Gray, Published by Microsoft Research in 2009. In it, it is argued that science has entered a completely new way of working, in which massive amounts of data are processed by efficient machine learning algorithm so as to help scientists to gain knowledge in problems otherwise intractable. Some authors even speak of the fifth paradigm, in which insights are gained by feeding artificial intelligence algorithms with synthetic data, coming from simulations.

All these concepts will be over viewed along this book, that could serve as a textbook for a post-graduate semester, typically consisting of some 60 h of lectures and laboratory assignments. The book is structured into four distinct parts. The first part is devoted mainly—but not only—to unsupervised machine learning techniques. In it, concepts such as the intrinsic dimensionality of a data set, the manifold hypothesis and manifold learning are described and analyzed with examples.

The second part of the book is devoted to techniques that we have grouped within the label of "learning". In other words, we review techniques—in general, supervised—that allow not only to gain insight about the information included in the data, but to make predictions in situations different that those for which they were trained.

Experimental data sets are very often extremely high dimensional. This makes it difficult to manage these data to gain insight, first, and to make predictions, which is the ultimate objective of this book. To overcome this difficulty, order reduction techniques have been developed in the last decades that allow us to simplify these data, by projecting them to lower-dimensional spaces. The third part of the book is devoted to these techniques. We will see how to develop reduced-order models once the intrinsic dimensionality of the data has been unveiled, with a fraction of the computational cost of employing brute data.

Finally, the fourth part of the book is devoted to the construction of digital twins, by leveraging all the just presented techniques. These twins make extensive use of data assimilation techniques, that constitute the main ingredient of this fourth part.

Open Access This chapter is licensed under the terms of the Creative Commons Attribution-NonCommercial-NoDerivatives 4.0 International License (http://creativecommons.org/licenses/by-nc-nd/4.0/), which permits any noncommercial use, sharing, distribution and reproduction in any medium or format, as long as you give appropriate credit to the original author(s) and the source, provide a link to the Creative Commons license and indicate if you modified the licensed material. You do not have permission under this license to share adapted material derived from this chapter or parts of it.

The images or other third party material in this chapter are included in the chapter's Creative Commons license, unless indicated otherwise in a credit line to the material. If material is not included in the chapter's Creative Commons license and your intended use is not permitted by statutory regulation or exceeds the permitted use, you will need to obtain permission directly from the copyright holder.

Chapter 2
Extended Summary

The 20th century engineering was mainly based on the use of models for optimally designing components and systems. In general, these models consist of mathematical operators able to convey the input towards its associated output. To be more concrete, imagine for a while a deformable solid whose shape is modified under the action of a system of forces applying on its surface. Such deformation entails an internal mechanical state in the solid, at the origin of reversible or irreversible geometrical changes, for example those occurring in manufacturing processes, as for instance in stamping or forging, just to cite a few.

Thus, the mechanical model referred above could be viewed as the operator, that from a given loading, infers the final mechanical state in the structural system.

There are two main typologies of models, referred later as Type I and Type II:

1. The one (Type I) that expresses the input/output relationship in an algebraic (linear or nonlinear) form. For example, a rod in which a tension applies, the applied stress is proportional to the induced deformation, the last defined as its length relative change. For small levels of deformation that relation remains linear, for larger deformations it can become nonlinear, and for even larger, it can become irreversible (inelasticity).

   These models are easily manipulable, and even when they compose in larger multi-component systems, their algebraic nature facilitates enormously its efficient manipulation and the solution of the associated problems.

2. The one (Type II) in which the input/output relationship results in a mathematical problem involving derivatives, in space and/or time. This occurs when the solution in a point and time depends on the past history and on the solution in all the other points in the considered domain. In that case models are expressed by more complex *mathematical objects* called partial differential equations.

Models of Type II, are, in general, more challenging from the point of view of its efficient solution. If we consider a one-dimensional model, a rod for instance, and we are interested in knowing the internal state in all the points in it, a first difficulty appears: there is an infinite number of unknowns, as many as points.

When the problem can be solved analytically, the difficulty disappears, with the solution closed form, the displacement in the present example, $u(x')$, expressed from the force applied everywhere, $f(x)$, that is, the displacement u at each point x' in the rod, results from the force f applied at each point x in the rod (for given boundary conditions).

However, most of models cannot be solved analytically. In general they are defined in too complex geometries Ω, with too complex boundary conditions and sometimes they are strongly nonlinear and coupled. For these reasons in the 20th century, discretization techniques became major protagonist of SBE (simulation based engineering), as for instance finite elements, finite differences, finite volumes to cite a few.

These techniques look for the solution at few points (the so-called nodes) and time instants, instead of looking for it at the infinite number of points and times. These procedures were called *discretization techniques*, and the finite element method became one of the most popular, today very well anchored in industrial practices.

However, from its conceptual formalization to the just mentioned industry conquer, the journey was rich. When solving our one-dimensional problem just referred, the chosen discretization technique, transforms the initial problem into the one that consists of calculating the solution in only N points (the so-called nodes) distributed on the rod, defining the so-called grid or mesh.

For the sake of illustration, we consider $N = 10$. Thus, as soon as the solution is calculated at these 10 points, it can be extended to any other location by simple interpolation. However, if we are in two dimensions (imagine a square domain), now the grid will contain 10×10 points (where the solution should be calculated) and in three dimensions (a cube for instance), $10 \times 10 \times 10$ points. Thus, the algebraic problems that result in the one-, two- and three-dimensional discretization, will have 10, 100 and 1000 unknowns respectively. The number of unknowns scales with the number of nodes in the grid or mesh, that as just mentioned grows exponentially with the dimension of the space where the problem is defined.

Even if an algebraic problem of size 10 can be solved at hand, the one involving the solution of a problem of size 1000, needs almost, the whole life of a human being.

Thus, discretization, compulsory for solving problems involving models of Type II, was only possible with the advenement and contribution of computers.

Even if a computer solves quickly algebraic system of equations, the just referred one involving 1000 unknowns will take some fractions of second in a standard laptop, engineering problems are becoming the more and more large and complex, and they generally involve hundreds of millions of unknowns, to be solved many times because of the nonlinearity of the problem or because of its transient nature (the solution evolves in time). In all these cases, even with the nowadays most powerful computational platforms, many simulations require days, weeks or even months.

In conclusion, most of models are solvable, with the help of powerful computers, but they need a consequent amount of time for performing that. The 20th century engineering was centered in components design, and for example energy, car or aerospace programs, were long enough to ensure the dialog between designers and simulation tools.

It is important to mention that design needs many simulations, to at least, partially explore the parametric domain, in order to find a quasi-optimal design. The cost of these simulations and the dimension of the parametric space (sometimes consisting of hundreds of parameters) make difficult, even today, optimal certified designs.

The other tricky point found in simulation-based engineering (SBE) concerns the quality of the model itself. Usual models come from centuries of fruitful science, that transformed the observed reality into mathematical objects, the so-called models. However, many aspects make difficult the real behavior apprehension and expression:

- The size of the analyzed system (in space and time) becomes the first enemy. A model could be considered globally good, but it could exhibit noticeable local deviations. In the same way, a model quite accurate at present, can reveal growing bias in time.
- The second difficulty lies in the large number of parameters that some models involve, with the associated difficulty of the proper calibration, that is, theirs parameters identification.
- Moreover, even when having the right model, the parameters exhibit variability. For example, the mechanical properties of specimens involving the same material, provided by the same supplier, produced by the same machine the same day, and elaborated in the same nominal conditions, exhibit slight differences. Thus, the model parameters distribute statistically, inducing an uncertainty that will propagate along the considered system.
- Finally, it remains an epistemic ignorance. The model and the reality can differ at some scale (in space and time) and then, the prediction to observation difference can grow as time advances.

In general engineers tried, in their practice, to limit the impact of the just referred issues that could compromise the predictions and with them, the quality of the designs.

2.1 From Virtual to Digital Twins

The 20th century engineering was prolific, with many impressive successes: aeronautics, space conquest, transport, energy, civil infrastructures, ...

In other domains, were models remained much less stablished (for instance in marketing, economy, social sciences, medicine, to cite a few), having fewer prognostic capabilities, scientists turned their view to the use of data, to make diagnosis and prognosis, and very quickly, major achievements were reported.

But then, three major and new facts irrupted in the fairly quiet world of engineering:

1. With the massively collected data and the communication technologies (IoT), classical offline engineering was pushed to operate online. The product-based engineering was replaced by a new performance-based engineering. Air transportation companies are not interested in aircraft engines anymore, but in hours of flight. You are not interested in buying an electric drill, but surely more in buying a god quality hole! Engineering does not stop when the car leaves the assembling chain, engineering should remain on board all along the car life, and for that, a main imperative: predicting fast and well.
2. Engineering is conquering larges spaces. In the past, engineering focused in the part, the component, the machine, …but now, engineering must address the connected world: smart industry, smart city, smart nation, …Thus, industry experienced its fourth revolution: after steam, electricity and electronic/automation, data joined the scene. Engineering moved from the part to the complex system of systems.
3. Moreover, the imminent next revolution (the fifth) will put definitively the human at the core: human-centric immersive engineering, in a physics-aware metaverse, that with the impressive progresses in computer vision, multimodal devices and NLP (natural language processing), with ChatGPT overwhelming all our preconceived ideas and vision, …seems defy the Turing test.

However, with the human integrated into the system, metaverse modeling becomes trickier, with the human escaping from, or resisting to, usual modeling approaches.

It was there, when engineering science turned its interest to the new possibilities that data and artificial intelligence offer, the expectations being quite large.

The learning process seems quite general and agnostic with respect to the considered physics. As soon as input and output data are collected, here referred by f and u respectively, $(f_i, u_i), i = 1, \ldots, D$ (with D the number of available data), one could try to create the relation between both, that is $u = \mathcal{F}(f)$.

As soon as the functional relation $\mathcal{F}(\cdot)$ is learnt and established, now, for any new input f, it suffices to apply the learnt model $\mathcal{F}(f)$, to obtain the associated and searched output u, $u = \mathcal{F}(f)$. Thus, we learn from data (machine learning), and as soon as the model is learnt, it can be applied in almost real-time to any other scenario.

In the previous discussion, all the challenge reduced to obtain the so-called regression $u = \mathcal{F}(f)$, that when applied on f, produces the associated response, u. There are many technologies for constructing such regressions, whose choice depends mainly on the quantity of available data (here noted by D), the time available for performing the training (online versus offline), and also on the data quality, among other factors.

In summary, if the quantity of interest on which we are interested (here u) has been well identified, as well as all the variables affecting it (here f), if the amount of data (here D) is large enough, as well as its quality (variability and precision), there are numerous techniques of machine learning able to extract the functional relationship (regression) $u = \mathcal{F}(f)$.

As just mentioned, data allows to extract models, under certain constraints on the amount of data, its quality, the processing time, etc. and the procedure seems quite general, and at some point, quite simple!

However, everything is not so simple, some remaining difficulties persist:

- Which data to collect, at which scale, when and where. The collected data will be useful, that is, it should contribute to the searched output. No feature contributing to the output must be discarded. Technologies exist to remove useless features and also to discover missing features, or even to operate in their absence.
- In engineering data has a cost. Big-data is not a valuable option in engineering. We cannot collect the data that we want and when we want. We collect the data that we can and when we can. Data is expensive from one side (sensing devices, instrumentation, data communication, data mining and processing), from the other, measurements are confronted to technological difficulties to put in place, without forgetting regulations. The environmental footprint is becoming also an essential criterion to consider.
- Even when succeeding to extract the regression $u = \mathcal{F}(f)$, explaining it remains more difficult. It is easy nowadays, after centuries of science, to explain the gravitation and its consequences on the trajectory of planets. However, explaining the data collected by the Babylonian astronomers remains at first view, less obvious, even if the predictions performed from them could be quite accurate.
- A model can be applied everywhere (within its domain of validity). In the case of data, when the extracted model is applied far from the region where the data that served to construct it was collected, extrapolation occurs, with the very well-known associated risks.

2.2 In Between Models and Data

As just discussed, the almost fully physics-based option has advantages but also drawbacks (computing time and computational resources, the difficulty of addressing very large systems involving noticeable variability and uncertainty).

On the opposite side, the almost fully data-driven option seems a valuable route when models do not exist or exhibit noticeable lack of accuracy. However, this route is also confronted to many difficulties, when applied in engineering practices, among them: the data cost, the extrapolation risk and the difficulty to explain, and then certify.

Perhaps the best choice consists of combining both settings instead of choosing between one or the other.

Among the numerous possibilities, here we consider two:

1. Physics-informing learning,
2. Physics-augmented learning.

In both cases, we make use of the fact that learning a part becomes, sometimes, cheaper that learning the whole. In physics, conservation laws are assumed being universal, and today it is not necessary to learn them again, but now from data. However, more phenomenological relationships can be and could be approached from a data-driven perspective, to improve them.

2.2.1 Physics-Informed Learning

Imagine for a while that we are looking for the output u at each point x in a domain Ω, i.e. $u(x)$, $x \in \Omega$. The regression expressing $u(x)$ is expected to be complex and strongly nonlinear. Here, instead of using the finite element method as previously mentioned, we decide to approximate $u(x)$ by using a nonlinear machine-learning-based regression, for instance a neural network –NN–.

On the other hand the stablished physics, here assumed applicable, states that $u(x)$ verifies a state-of-the-art model, expressed from a partial differential equation, expressed in the generic form: $\mathcal{L}(u(x)) = f(x)$, where $\mathcal{L}(\cdot)$ represents a linear or nonlinear differential operator.

The PINN (physics informed neural network) looks for the neural network expressing u from x, that is, approximating $u(x)$, in such a way that it verifies its governing equation, that is, it minimizes the residual $\mathcal{R}$, with $\mathcal{R} = \|\mathcal{L}(u(x)) - f(x)\|$, that forces during the construction of the regression $u(x)$ the verification of the partial differential equation. Automatic differentiation allows taking derivatives of NN-based regressions to enforce the partial differential equation fulfillment.

PINN can be viewed as a sort of collocation method, with an approximation of the unknown field, here $u(x)$, given by a NN, and a constructor based on the residual minimization. It allows easily assimilating data in a quite simple and transparent way.

2.2.2 Physics-Augmented Learning

Here, the main idea consists of assuming that the reference solution $u(x)$, that results from a given loading term $f(x)$, consists of two contributions. The first describes all our knowledge, that is, the component that comes from the physics-based model, here noted by $u^P(x)$.

Now, the gap between real behavior $u(x)$ and the physics-based model prediction $u^P(x)$ is computed: $u(x) - u^P(x)$. This gap is usually called *ignorance*. Now, machine learning is used to model that gap, from which it results the so-called data-driven model: $u^D(x)$. Thus, the fundamental approximation reads: $u(x) = u^P(x) + u^D(x)$.

This relation is at the origin of the so-called hybrid modeling paradigm or the hybrid twin when applied to a particular system or asset.

2.2.3 Discussion

Both approaches allow reducing drastically the amount of data needed for constructing the models, because of the provided information or because of the fact that one only looks for the model of the gap, that is assumed much simpler to approximate than the real behavior itself.

Moreover, the use of the physics-based model allows to explain at least the most significant part of the model foundations and its derived predictions, alleviating at the same time extrapolation issues.

The reduction on the amount of data is accomplished not only because of the use of the physics, but also because that physics informs on the optimal location and time to collect the data. It is the case when considering the so-called *active learning*, that becomes specially performant when making use of the existing physics-based knowledge.

For the sake of illustration we consider the fact of accessing to the temperature in Paris. To apprehend the temperature in Paris one does not need to place thousands on thermometers in each street and collecting the temperature each millisecond. Knowing the physics, that shapes our experience and common sense, one knows that placing few thermometers (one in each district for example) and taking the temperatures in the morning, at noon, in the afternoon and at mid-night, could largely suffice.

In a general way, we should take profit of all the existing knowledge.

2.2.4 The Global Picture

The just referred discussion is summarized in Fig. 2.1, that illustrates the three main paradigms: (i) model-based; (ii) data-driven; and (iii) the informed and/or augmented one.

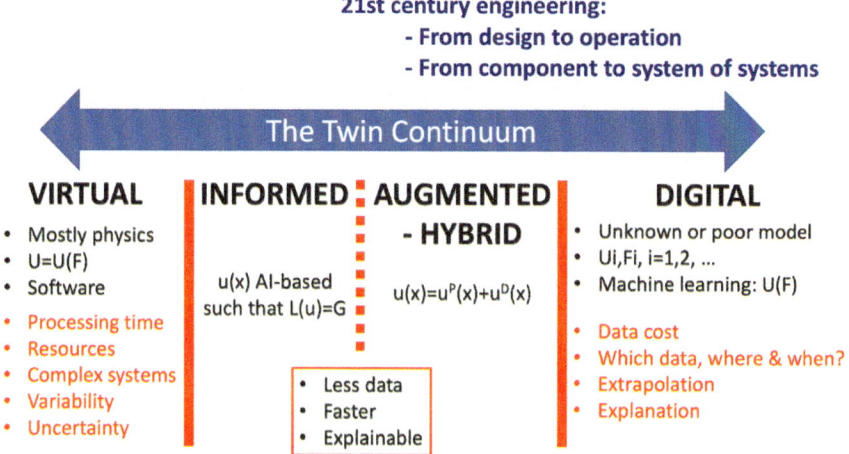

Fig. 2.1 The twin continuum

2.3 Physics in Real-Time: Model Order Reduction

A first need concerns the Physics-based model, whose solution procedure must be accelerated to ensure real-time performances.

For this purpose, we consider the construction of a surrogate, that is, the solution of the physics-based model for any choice of the parameters that it involves. This surrogate, once constructed, can infer the solution almost in real-time, as soon as the model parameters are provided.

Parametric solutions enable simulation, optimization, inverse analysis, simulation-based control and uncertainty propagation, all them under the stringent real-time constraint. These parametric solutions make possible model-based real-time engineering, in an extremely efficient and unprecedented manner.

For the sake of illustration, we consider a simple parametric solution expressed by $u(x; p, q)$, where the unknown u depends at each position x, on two parameters, here noted by p and q. Here, for the sake of simplicity we consider only two parameters, but considering more is straightforward.

The surrogate is constructed as follows:

1. A Design of Experiments (DoE) is defined, by using some sampling strategy. Among the numerous possibilities, Latin Hypercube, quadrature (e.g. Gauss-Lobatto-Chebyshev or Smolyak), or those associated with active learning (gaussian processes, Fisher information matrix, to cite a few) are widely considered, to provide the space sampling: $(p_1, q_1), \ldots, (p_D, q_D)$.
2. Then, by using the physics-based model at hand, and more particularly a software able to solve it, the solution associated to each point of the DoE is calculated: $u_1(x) = u(x; p_1, q_1), \ldots, u_D(x) = u(x; p_D, q_D)$.
3. Data consisting of those high-fidelity solutions, can be reduced by using linear or nonlinear dimensionality reduction, to facilitate the subsequent regressions.
4. Finally, a regression $u(x; p, q)$ is constructed, such that the loss (the approximation error) is minimized. The loss L is usually defined from the L2-norm of the error, i.e. $L = \sum_{i=1}^{D} \|u(x, p_i, q_i) - u_i(x)\|_2$. Different procesures exist, consisting of nonlinear or regularized linear regressions.

2.4 Data and Learning

Regressions, able to extract models from data, are of different nature, depending on the nature of the manipulated data.

The most usual typologies of data are:

- Lists (tables), involving continuous or discrete numerical values, categorial features, ...;
- Images (real 2D or 3D images or numerical simulation results);
- Tensor formats, describing compressed images or data;

2.4 Data and Learning

- Graphs;
- Curves;
- Time-series, similar to curves, but fulfilling causality.

Among the most usual machine learning techniques, we can mention:

- Regularized polynomial regressions, particularly appealing when data is scarce and the solution expressible from usual or engineered (making use of existing knowledge) polynomial bases.
- Artificial Neural Networks (NN) that perform very well as soon as the amount of data is large enough, and the different hyper-parameters well tuned. The universal approximation theorem states that any function can be approximate by using a single layer neural network with an adequate number of neurons. This results offers to NN an enormous versatility and power.
- Auto-Encoder –AE– is a NN architecture composed of a coding and decoding units, that maps data into a latent space that approximate the intrinsic data dimensionality. Modeling operates better at the latent level where most of the linear and nonlinear correlations have been removed. Other data-driven linear and nonlinear dimensionality reduction techniques are based on manifold learning, as PCA, kPCA, LLE, tSNE, ...
- GAN, generative adversarial networks, improve robustness with respect to the noise and provide a valuable data augmentation.
- CNN, convolutional neural networks, are specially adapted to address images, to recognize and extract particular patterns. Convolution operating on an unstructured mesh, for instance a graph, leads to the so-called GNN (graph neural network) with appealing properties for learning primal and dual physical behaviors. GNN are very close to usual discretization techniques, but GNN learn the subjacent models instead of operating on the know models for solving them.
- When considering time series, rNN (recurrent NN) and LSTM (long short time memory) are specially performant, with a subtle link between memory, the hidden physics and its associated missing data.
- Dynamical systems are usually addressed with different technologies and approaches, the ones just mentioned (rNN and LSTM) that learn the state evolution and its associated observables, ResNet or NeuralODE that learn the dynamical system forcing term, or by using DMD (dynamic model decomposition) or the associated Koopman operator that allows to address nonlinear behaviors in a more efficient way. The nowadays very popular *reservoir computing* and DeepONets seem plenty of potentiality.
- When moving to the domain of informed / augmented learning, the more usual techniques are the PINN (physics informed neural network), the TINN (thermodynamic informed NN) also called SPNN (structure preserving NN), or the ones making use of the model-based / data-driven hybridation. The last can be viewed as a sort of transfer learning.

In what concerns the learning modalities, we can cite: supervised, unsupervised, self-supervised and semi-supervised learning, where different amount of information or knowledge is used during the learning process. Transfer learning is also a valuable route for transferring knowledge from one domain to another. The learning process can be also enhanced by using reinforcement, specially useful in stochastic settings.

As previously mentioned, there is not a unique or universal choice. The choice depends on the amount of data, the expected complexity of the manifold defined by the data, or the necessity of operating (and eventually also learning) online.

2.5 Hybridation

Hybridation can follow two different routes, the one that consists of enriching the solution and the one that enrich the model, from which the enriched solution will result.

2.5.1 Solution Enrichment

With the physics-based model solution given $u^P(x, \mathbf{p})$ (computed by using the techniques previously mentioned), where boldface $\mathbf{p}$ refers to the vector of parameters (if more than one), and with the data-driven correction, also learnt by using one of the technologies just referred, $u^D(x, \mathbf{p})$, the enriched (or corrected) solution reads $u^P(x, \mathbf{p}) + u^D(x, \mathbf{p})$.

Thus, online, as soon as data is collected, the model is calibrated, that is, the parameters grouped in $\mathbf{p}$ are identified, being noted by $\mathbf{p}^*$, by minimizing the difference between the model predictions and the measurements.

Then, the data-driven model is particularized for the just identified values of the parameters, and then the hybrid prediction is obtained from $u^P(x, \mathbf{p}^*) + u^D(x, \mathbf{p}^*)$. The same procedure applies in the case of transient problems.

2.5.2 Model Enrichment

When considering the model enrichment route, a nominal model is assumed known, whose discrete form (linear for the sake of simplicity) reads $\mathbf{MU} = \mathbf{F}$, where $\mathbf{F}$ and $\mathbf{U}$ are the nodal vectors of actions and reactions, i.e. respectively forces and displacements when considering an elastic problem.

Now, the displacement is collected at some locations and grouped into the vector $\tilde{\mathbf{U}}^{\text{exp}}$. At these locations, the nominal model predicts the displacements $\tilde{\mathbf{U}}$, and a noticeable gap is noticed, that is, the norm of the displacements difference $\|\tilde{\mathbf{U}}^{\text{exp}} - \tilde{\mathbf{U}}\|$ remains greater than the assumable error ϵ, i.e. $\|\tilde{\mathbf{U}}^{\text{exp}} - \tilde{\mathbf{U}}\| > \epsilon$.

Thus, it seems that the model expressed by the matrix $\mathbf{M}$ remains unable to represent the measurement, and that consequently, it needs an enrichment $\mathbf{M}^*$, such

2.5 Hybridation

that the enriched discrete model $\mathbf{M} + \mathbf{M}^*$ should represent the collected data, that is, the corrected model $\mathbf{M} + \mathbf{M}^*$ and the associated displacement prediction $\mathbf{U} + \mathbf{U}^*$, should verify the equilibrium, i.e. $(\mathbf{M} + \mathbf{M}^*)(\mathbf{U} + \mathbf{U}^*) = \mathbf{F}$ while representing the collected data, i.e. $\|\tilde{\mathbf{U}}^{\text{exp}} - (\tilde{\mathbf{U}} + \tilde{\mathbf{U}}^*)\| < \epsilon$.

Now, a parametrization of the model correction $\mathbf{M}^*$, combined with an adequate regularization, could serve to compute the model enrichment $\mathbf{M}^*$ while performing the data completion $\mathbf{U} + \mathbf{U}^*$.

When following this route, the main issues concern: (i) the choice of the functional to be minimized for performing the hybridation; (ii) the enrichment parametrization (reduced basis, matrix parametrization, ...); and (iii) ensuring the properties on the enriched model (e.g. stability when addressing transient simulations, positivity, to cite few).

Open Access This chapter is licensed under the terms of the Creative Commons Attribution-NonCommercial-NoDerivatives 4.0 International License (http://creativecommons.org/licenses/by-nc-nd/4.0/), which permits any noncommercial use, sharing, distribution and reproduction in any medium or format, as long as you give appropriate credit to the original author(s) and the source, provide a link to the Creative Commons license and indicate if you modified the licensed material. You do not have permission under this license to share adapted material derived from this chapter or parts of it.

The images or other third party material in this chapter are included in the chapter's Creative Commons license, unless indicated otherwise in a credit line to the material. If material is not included in the chapter's Creative Commons license and your intended use is not permitted by statutory regulation or exceeds the permitted use, you will need to obtain permission directly from the copyright holder.

Part I
Around Data

Chapter 3
Intrinsic Dimensionality of a Data Set and Manifold Learning

We begin this journey by reviewing some techniques able to extract the intrinsic dimensionality of a data set. These techniques have been employed successfully in different engineering applications and have revealed to be of an utmost importance in the treatment of big data sets [1–3].

Let $\mathbf{y} \in \mathbb{R}^D$ be a vector that contains the available data (extracted from experiments or coming from numerical simulations). These data are represented as points in a space of dimension D (see Fig. 3.1). However, correlations exist among these data, so that they are expected distributing on a low-dimensional subspace embedded into $\mathbb{R}^D$. Manifold learning techniques try to extract the intrinsic dimensionality d (in general much smaller than D, i.e., $d \ll D$) of the reduced subspaces, as shown in Fig. 3.1. The fact that high-dimensional data arising from physical phenomena are usually embedded onto low-dimensional manifolds receives the name of the *manifold hypothesis*, something that is widely accepted in many branches of science.

Let us assume that we have performed some measurements on our physical system, so that we have extracted M different snapshots $\mathbf{y}_1, \ldots, \mathbf{y}_M$ of the system. These snapshots are organized in the columns of a $D \times M$ matrix $\mathbf{Y}$. The corresponding $d \times M$ reduced matrix Ξ is composed of the associated vectors $\boldsymbol{\xi}_i, i = 1, \ldots, M$.

The next sections revisit some widely employed manifold learning techniques [4], which will be analyzed in detail throughout this monograph.

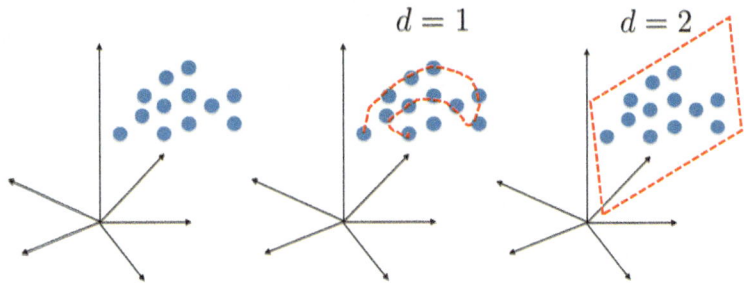

Fig. 3.1 Data points (left), one-dimensional manifold (center) and two-dimensional (without loss of generality depicted linear) manifold (right)

3.1 Principal Component Analysis, PCA, and Its Local Counterpart, ℓPCA

PCA employs a linear transformation $\mathbf{W}$ verifying $\mathbf{W}^T \mathbf{W} = \mathbf{I}_d$—with $\mathbf{I}_d$ the identity matrix of size d—, such that $\mathbf{y} = \mathbf{W}\boldsymbol{\xi}$. Data $\mathbf{y}_i$ and $\boldsymbol{\xi}_i$, $1, \ldots, M$ (both assumed centered) define the columns of matrices $\mathbf{Y}$ and $\boldsymbol{\Xi}$ respectively. PCA looks for a maximal variance and decorrelation in the latent variable set $\boldsymbol{\xi}$.

The covariance matrix $\mathbf{C}_{yy} = \mathbb{E}\{\mathbf{YY}^T\}$ (where $\mathbb{E}$ refers to the statistical expectation) can be factorized, for example by employing spectral decomposition, $\mathbf{C}_{yy} = \mathbf{V}\boldsymbol{\Lambda}\mathbf{V}^T$.

Thus, $\mathbb{E}\{\mathbf{YY}^T\} = \mathbb{E}\{\mathbf{W}\boldsymbol{\Xi}\boldsymbol{\Xi}^T\mathbf{W}^T\} = \mathbf{W}\mathbb{E}\{\boldsymbol{\Xi}\boldsymbol{\Xi}^T\}\mathbf{W}^T$. By pre- and post-multiplying by $\mathbf{W}^T$ and $\mathbf{W}$ respectively, it results $\mathbf{W}^T\mathbb{E}\{\mathbf{YY}^T\}\mathbf{W} = \mathbb{E}\{\boldsymbol{\Xi}\boldsymbol{\Xi}^T\}$ or $\mathbf{W}^T\mathbf{V}\boldsymbol{\Lambda}\mathbf{V}^T\mathbf{W} = \mathbb{E}\{\boldsymbol{\Xi}\boldsymbol{\Xi}^T\}$, where the attended diagonal character of $\mathbb{E}\{\boldsymbol{\Xi}\boldsymbol{\Xi}^T\}$ results from the choice $\mathbf{W} = \mathbf{V}\mathbf{I}_{D\times d}$ (with $\mathbf{I}_{D\times d}$ the $D \times d$ unit matrix), i.e., $\mathbf{W}$ is composed by the d eigenvectors associated with the d non-zero eigenvalues of matrix $\mathbf{C}_{yy}$, that represent the variance of the latent variables $\boldsymbol{\xi}$.

PCA is essentially a linear technique, so that it works specially well in the case of linear manifolds. However, as sketched in Fig. 3.2 it fails when the manifold exhibits significant nonlinearities.

Fig. 3.2 Intuitive representations of PCA (left) and its limitation for representing nonlinear manifolds (centre). Even if the red point seems very close to the blue one just in front of it, it is in fact far when computing distances on the manifold, the only that have a real sense. In that case applying PCA locally (right) allows a better extraction of the local intrinsic dimensionality

To avoid this, the local-PCA, ℓPCA, extracts the local dimensionality by applying locally the just described PCA technique. By invoking the essential characteristic of a manifold, its equivalence to a flat space in the neighborhood of each point, Local Tangent Space Alignment (LTSA) unveils the geometry of the manifold locally by constructing an approximation to the tangent space at each data point [5].

3.2 Multidimensional Scaling, MDS

While PCA employs the covariance matrix of the data, $\mathbf{YY}^T$, the multidimensional scaling works with the so-called Gram matrix. It contains scalar products of the snapshots, $\mathbf{S} = \mathbf{Y}^T \mathbf{Y}$, whose decomposition reads $\mathbf{U\Lambda U}^T$.

Taking into account $\mathbf{Y} = \mathbf{W\Xi}$ and $\mathbf{W}^T\mathbf{W} = \mathbf{I}$, it results $\mathbf{Y}^T\mathbf{Y} = \mathbf{\Xi}^T\mathbf{\Xi}$, and consequently $\mathbf{\Xi} = \mathbf{I}_{d\times M}\mathbf{\Lambda}^{1/2}\mathbf{U}^T$.

3.3 Kernel Principal Component Analysis, kPCA

Kernel Principal Component Analysis, kPCA, aims at mapping the data into a space of adequate dimensionality Q, that could be even much higher ($Q \gg D$), such that the initially nonlinear manifold in $\mathbb{R}^D$ becomes linear in $\mathbb{R}^Q$, making thus possible the application of the standard PCA.

With $\mathbf{Z}$ representing the images $\mathbf{z}_i \in \mathbb{R}^Q$ of the data-points $\mathbf{y}_i \in \mathbb{R}^D$, PCA can be applied without the need of defining explicitly that mapping $\mathbf{y} \to \mathbf{z}$, because to apply PCA only a scalar product in $\mathbb{R}^Q$ is needed. That is, the application of PCA in $\mathbb{R}^Q$ only needs the use of a scalar product in that space, and the Mercer theorem allows computing a scalar product in $\mathbb{R}^Q$ while operating in the initial space $\mathbb{R}^D$.

If $\mathbf{z}_i$ is assumed centered, then the resulting eigenproblem reads

$$\frac{1}{M}\mathbf{ZZ}^T\mathbf{v}_j = \lambda_j\mathbf{v}_j,$$

that with each eigenvector expressible from the existing data, that is, $\mathbf{v}_j = \mathbf{Z}\boldsymbol{\alpha}_j$, it results

$$\frac{1}{M}\mathbf{ZZ}^T\mathbf{Z}\boldsymbol{\alpha}_j = \lambda_j\mathbf{Z}\boldsymbol{\alpha}_j. \tag{3.1}$$

As previously indicated, the scalar product in $\mathbb{R}^Q$ is calculated in $\mathbb{R}^D$ by invoking the so-called Mercer theorem, and the so-called *kernel trick*: $\kappa(\mathbf{y}_i, \mathbf{y}_j) = \mathbf{z}_i \cdot \mathbf{z}_j$.

Thus, $\mathbf{Z}^T\mathbf{Z}$ is calculated without the need of knowing the explicit form of $\mathbf{Z}$, from the use of the kernel trick, that allows expressing

$$\mathbf{Z}^T\mathbf{Z} = \mathbf{K}, \tag{3.2}$$

with $K_{ij} = \kappa(\mathbf{y}_i, \mathbf{y}_j)$.

Thus, by premultiplying Eq. (3.1) by $\mathbf{Z}^T$ and considering Eq. (3.2), it results $\mathbf{K}^2\alpha_j = M\lambda_j \mathbf{K}\alpha_j$, or

$$\mathbf{K}\alpha_j = M\lambda_j \alpha_j, \qquad (3.3)$$

that enables performing the linear reduction by considering the d non-zero (or the d highest) eigenvalues. Centering $\mathbf{Z}$ can be performed operating directly on $\mathbf{K}$ [4]. The d eigenvectors $\mathbf{v}_i$ compose the columns of a matrix $\mathbf{V}$ whereas the reduced eigenvectors α_i are stored in the columns of a matrix $\mathbf{A}$, with $\mathbf{V} = \mathbf{ZA}$.

Now, when considering a new data $\mathbf{y}$, it suffices to project its image $\mathbf{z}$ on the d eigenvectors $\mathbf{v}_j$, i.e. $\boldsymbol{\xi} = \mathbf{V}^T \mathbf{z}$, or $\boldsymbol{\xi} = \mathbf{A}^T \mathbf{Z}^T \mathbf{z}$, where the product of each row of $\mathbf{Z}^T$ and the column vector $\mathbf{z}$ makes use of the kernel trick.

It is important no note that $\boldsymbol{\xi}$, is defined in the space $\mathbb{R}^d$, but that in the present case, when reducing from Q to d, d could be greater, smaller or equal to D. When $d < D$ the kPCA acts as a nonlinear dimensionality reduction technique, whereas, when $d > D$, the kPCA is informing us on the lack of completeness of data $\mathbf{y}$, as discussed and visualized in [6].

3.4 Locally Linear Embedding, LLE

Starting again from our data points $\mathbf{y}_i \in \mathbb{R}^D$, $i = 1, \ldots, M$, Locally Linear Embedding, LLE, methods operate in two steps [7]:

1. LLE first computes a linear interpolation of each data point $\mathbf{y}_i$, $i = 1, \ldots, M$ from its surrounding neighbors. The analyst must first choose a number K of nearest neighbors: $\mathbf{y}_i = \sum_{j \in \mathcal{S}_i} W_{ij} \mathbf{y}_j$ (with $\mathcal{S}_i$ the set of the K-nearest neighbors of $\mathbf{y}_i$). This is done often by trial and error, or by invoking the intuition of the analyst, but it is frequently very easy to find the appropriate number of neighbors for an optimal result. LLE then computes the weights W_{ij} by minimizing the functional $\mathcal{F}(\mathbf{W}) = \sum_{i=1}^{M} \|\mathbf{y}_i - \sum_{j \in \mathcal{S}_i} W_{ij} \mathbf{y}_j\|^2$.
2. Each patch of neighboring data points around $\mathbf{y}_i$, $\forall i$, is mapped onto a lower dimensional embedding space of dimension $d \ll D$ keeping the just computed interpolation weights. The method must now find the coordinates $\boldsymbol{\xi}_i \in \mathbb{R}^d$ (related to $\mathbf{y}_i$) that satisfy the same neighboring relationship of the high-dimensional space, by minimizing a second functional: $\mathcal{G}(\boldsymbol{\xi}_1, \ldots, \boldsymbol{\xi}_M) = \sum_{i=1}^{M} \|\boldsymbol{\xi}_i - \sum_{j \in \mathcal{S}_i} W_{ij} \boldsymbol{\xi}_j\|^2$.

3.5 t-Distributed Stochastic Neighbor Embedding, tSNE

t-distributed Stochastic Neighbor Embedding [8] is not actually a technique aiming at extracting the intrinsic dimension of the data manifold, but a technique to visualize high-dimensional data by mapping them onto a manifold of low dimension—very often only two dimensions—so as to enable their visualization on a computer screen.

The main issue of this method is the choice of the mapping in order to retain as many valuable information and data relations as the ones existing in the high-dimensional space. tSNE is based on the Stochastic Neighbor Embedding, SNE.

3.5.1 Stochastic Neighbor Embedding, SNE

In SNE, the similarity (or proximity) of data in the original high-dimensional space, $\mathbf{y}_i$ and $\mathbf{y}_j$, both in $\mathbb{R}^D$, is evaluated from the conditional probability

$$P_{\mathbf{y}_j|\mathbf{y}_i} = \frac{\exp\left(-\|\mathbf{y}_i - \mathbf{y}_j\|^2/2\sigma_i^2\right)}{\sum_{k \neq i} \exp\left(-\|\mathbf{y}_i - \mathbf{y}_k\|^2/2\sigma_i^2\right)}$$

and a similar expression for the probability in the reduced space $Q_{\xi_j|\xi_i}$, using again a gaussian function but now enforcing the scaling by assuming all the variances equal to $1/\sqrt{2}$.

The mapping is constructed by enforcing the proximity of probability distributions P_i and Q_i, and for that, the cost function $\mathcal{C}$ employed is based on the use of the Kullback-Leibler, KL, divergence, i.e.

$$\mathcal{C} = \sum_i KL(P_i \| Q_i) = \sum_i \sum_j P_{\mathbf{y}_j|\mathbf{y}_i} \log \frac{P_{\mathbf{y}_j|\mathbf{y}_i}}{Q_{\xi_j|\xi_i}},$$

that results in

$$\frac{\delta \mathcal{C}}{\delta \xi_i} = 2 \sum_j \left(P_{\mathbf{y}_j|\mathbf{y}_i} - Q_{\xi_j|\xi_i} + P_{\mathbf{y}_i|\mathbf{y}_j} - Q_{\xi_i|\xi_j} \right) (\xi_i - \xi_j),$$

which represents a sort of spring that attracts or repels depending on the probability distributions mismatch.

The variances in the high-dimensional space are computed to ensure a value of the perplexity $\mathcal{P}_e$ defined by the user. The perplexity is defined by $\mathcal{P}_e = 2^{H(P_i)}$, with $H(P_i)$ the Shannon entropy $H(P_i) = -\sum_j P_{\mathbf{y}_j|\mathbf{y}_i} \log_2 P_{\mathbf{y}_j|\mathbf{y}_i}$.

The main drawbacks of SNE is its lack of symmetry (KL-divergences are not symmetric) and the complexity of computing reduced data points ξ_i, $\forall i$, by minimizing the cost function $\mathcal{C}$.

3.5.2 Coming Back to the tSNE

tSNE represents a sort of symmetric SNE where, instead of the conditional probability, it is the joint probability that is considered, i.e.,

$$C = \sum_i KL(P_i \| Q_i) = \sum_i \sum_j P_{y_j|y_i} \log \frac{P_{y_j|y_i}}{Q_{\xi_j|\xi_i}}.$$

In addition, the gaussian is used in the high-dimensional space but a t-Student is considered in the low dimensional space

$$Q_{\xi_i|\xi_j} = \frac{\left(1 - \|\mathbf{y}_i - \mathbf{y}_j\|^2\right)^{-1}}{\sum_{k \neq i} \left(1 - \|\mathbf{y}_i - \mathbf{y}_k\|^2\right)^{-1}}.$$

The heaver tails of the *t*-Student better represent distances by probabilities. This is important because of the so-called *crowding problem* related to the fact of accommodating enormous amounts of high-dimensional data within a limited region in the low-dimensional space.

3.6 PCA on Qualitative Variables

PCA can also operate on qualitative variables. If we consider D data, described with C coordinates $c = 1, \ldots, C$, each of them having V_c values, the data matrix will consists of a matrix $\mathbf{X}$ with $D \times V$ components, with $V = \sum_{c=1}^{C} V_c$.

The component $X(i, j) = 1$, $i = 1, \ldots, D$, and $j = 1, \ldots, V$, if the datum i contains the value j, and $X(i, j) = 0$ in the opposite case.

We define first

$$\mathbf{Z} = \frac{\mathbf{X}}{D \cdot V},$$

then, we consider vectors $\mathbf{R}$ and $\mathbf{C}$ that contains the sum of the components of $\mathbf{Z}$ by columns (to obtain the row vector $\mathbf{R}$) and by rows to obtain the column vector $\mathbf{C}$, and then define the diagonal matrices $\mathbf{D}_r = \mathtt{diag}(\mathbf{R})$ and $\mathbf{D}_c = \mathtt{diag}(\mathbf{C})$.

Then we can define the matrix $\mathbf{M}$

$$\mathbf{M} = \mathbf{D}_r^{-1/2}(\mathbf{Z} - \mathbf{R} \otimes \mathbf{C})\mathbf{D}_c^{-1/2},$$

whose SVD leads to matrices $\mathbf{P}$, Δ and $\mathbf{Q}$.

Thus,

$$\mathbf{F} = \mathbf{D}_r^{-1/2}\mathbf{P}\Delta,$$

and
$$\mathbf{G} = \mathbf{D}_c^{-1/2} \mathbf{Q} \Delta,$$

give the coordinates of the datum or the variable in the so-called factor space, respectively [9, 10].

3.7 PCA in Presence of Missing Data

There exist some techniques for minimizing the effect of missing values in multivariate data analysis, perform principal component analysis on incomplete data sets, aiming to obtain scores and graphical representations despite of the existence of missing values [11]. Other techniques able to proceed with missing data will be described later, in particular the so-called gappy-POD in Chap. 23.

References

1. D. Gonzalez, J.V. Aguado, E. Cueto, E. Abisset-Chavanne, F. Chinesta, kPCA-based parametric solutions within the PGD framework. Arch. Comput. Methods Eng. **25**, 69–86 (2018)
2. E. Lopez, D. Gonzalez, J.V. Aguado, E. Abisset-Chavanne, E. Cueto, C. Binetruy, F. Chinesta, A manifold learning approach for integrated computational materials engineering. Arch. Comput. Methods Eng. **25**, 59–68 (2018)
3. A. Badias, S. Curtit, D. Gonzalez, I. Alfaro, F. Chinesta, E. Cueto, An augmented reality platform for interactive aerodynamic design and analysis. Int. J. Numer. Methods Eng. **120**(1), 125–138 (2019)
4. J.A. Lee, M. Verleysen, *Nonlinear Dimensionality Reduction* (Springer, New York, 2007)
5. Z. Zhang, H. Zha, Principal manifolds and nonlinear dimensionality reduction via tangent space alignment. SIAM J. Sci. Comp. **26**(1), 313–338 (2004)
6. R. Ibanez, P. Gilormini, E. Cueto, F. Chinesta, Numerical experiments on unsupervised manifold learning applied to mechanical modeling of materials and structures. CRAS Mec. **348**(10–11), 937–958 (2020)
7. T. Roweis, L.K. Saul, Nonlinear dimensionality reduction by locally linear embedding. Science **290**, 2323–2326 (2000)
8. L. Maaten, G. Hinton, Visualizing data using t-SNE. J. Mach. Learn. Res. **9**, 2579–2605 (2008)
9. B. Escofier, Traitement simultane de variables quantitatives et qualitatives en analyse factorielle. Les Cahiers de Analyse des Donnees **4**(2), 137–146 (1979)
10. B. Escofier, J. Pages, *Analyses Factorielles Simples et Multiples* (Dunod, 2008)
11. F. Husson, J. Josse, missMDA: handling missing values with/in multivariate data analysis (Principal Component Methods). R package version 1.10 (2016). https://CRAN.R-project.org/package=missMDA

Open Access This chapter is licensed under the terms of the Creative Commons Attribution-NonCommercial-NoDerivatives 4.0 International License (http://creativecommons.org/licenses/by-nc-nd/4.0/), which permits any noncommercial use, sharing, distribution and reproduction in any medium or format, as long as you give appropriate credit to the original author(s) and the source, provide a link to the Creative Commons license and indicate if you modified the licensed material. You do not have permission under this license to share adapted material derived from this chapter or parts of it.

The images or other third party material in this chapter are included in the chapter's Creative Commons license, unless indicated otherwise in a credit line to the material. If material is not included in the chapter's Creative Commons license and your intended use is not permitted by statutory regulation or exceeds the permitted use, you will need to obtain permission directly from the copyright holder.

Chapter 4
Intrinsic Dimensionality and Autoencoders

As indicated, manifold learning aims at extracting the low-dimensional manifold in which high-dimensional data are embedded, reducing the dimensionality, and consequently the number of coordinates required to express the data. However, in the techniques just described the inverse mapping, that is, the high-dimensional data reconstruction from the reduced data, is a tricky issue.

To circumvent de just referred difficulty related to the inverse mapping, while performing on strongly nonlinear manifolds, the so-called autoencoders, AE, [1–3] represent a very valuable alternative.

AE employ neural networks, NN, to perform first data reduction (coding), and then performing the high-dimensional data reconstruction (decoding). The process is sketched in Fig. 4.1. An introduction to neural networks is given in Chap. 20.

The high-dimensional data $\mathbf{y} \in \mathbb{R}^D$ is mapped into the so-called latent space, consisting of a hidden layer composed of d neurons, in general $d \ll D$, leading to $\boldsymbol{\xi} \in \mathbb{R}^d$, the reduced expression of $\mathbf{y}$. Then, that reduced data $\boldsymbol{\xi}$ expands to the output layer, to recover $\mathbf{y}$.

A NN is trained to accomplish accurately both tasks, the coding and the decoding, while trying to keep as thin as possible the hidden layer, whose size approximates the data intrinsic dimensionality. That reduction becomes of major interest in data reduction, in denoising or in physics-aware machine learning [4].

The encoder q_ϕ maps $\mathbf{y} \to \boldsymbol{\xi}$, while the decoder p_θ maps $\boldsymbol{\xi} \to \mathbf{y}$. Thus autoencoding looks for

$$p_\theta, \ q_\phi = \mathrm{argmin}_{p_\theta^*, \ q_\phi^*} \|\mathbf{y} - (p_\theta^* \circ q_\phi^*)\mathbf{y}\|^2.$$

A simple coding (without hidden layers) reads $\boldsymbol{\xi} = \sigma(\mathbf{Wy} + \mathbf{b})$, with the decoding expressed from $\mathbf{y}' = \sigma'(\mathbf{W}'\boldsymbol{\xi} + \mathbf{b}')$, that combined results in minimizing the loss $\mathcal{L}(\mathbf{y}, \mathbf{y}')$

$$\mathcal{L}(\mathbf{y}, \mathbf{y}') = \|\mathbf{y} - \mathbf{y}'\|^2 = \|\mathbf{y} - \sigma'(\mathbf{W}'(\sigma(\mathbf{Wy} + \mathbf{b})) + \mathbf{b}')\|^2.$$

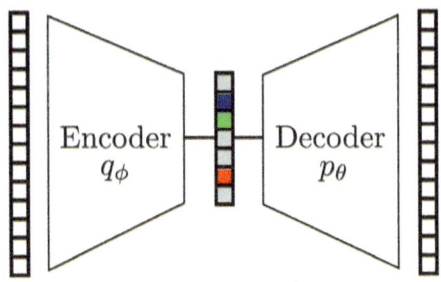

Fig. 4.1 Autoencoder composed of an encoding and a decoding process

By using the available data, the NN parameters ($\mathbf{W}$, $\mathbf{W}'$, $\mathbf{b}$, $\mathbf{b}'$) are calculated by using adequate minimization techniques.

To increase efficiency, avoid overfitting, and reduce the amount of data employed in training, different regularizations were proposed, giving rise to the so-called sparse autoencoders and other variants (variational, denoising, and contractive among many other choices) [1, 5, 6].

4.1 Sparse Autoencoders

In sparse autoencoders, SAE, the hidden layer (accommodating the reduced data) is larger, but sparsity is enforced though a penalty regularization $\mathbb{P}(\boldsymbol{\xi})$. Thus, the function to be minimized now reads $\mathcal{L}(\mathbf{y}, \mathbf{y}') + \mathbb{P}(\boldsymbol{\xi})$.

The most usual penalty formulations are based on:

- The Kullback-Leibler, KL, divergence.

  Assuming $\boldsymbol{\xi} \in \mathbb{R}^d$ and M data points available, we define the vector $\boldsymbol{\mu} \in \mathbb{R}^d$, such that each of its components μ_i is given by the local average

$$\mu_i = \frac{1}{M} \sum_{j=1}^{M} |\xi_i(\mathbf{y}_j)|,$$

  that is enforced to approach the value s, called sparsity parameter, close to zero.

  The penalty is formulated by using the KL divergence between a Bernoulli random variable with mean s and a Bernoulli random variable with mean μ_i, i.e.

$$\mathbb{P}(\boldsymbol{\xi}) = \sum_{i=1}^{d} KL(s||\mu_i).$$

- The ℓ^1–norm.

In this case we consider as penalty $\mathbb{P}(\boldsymbol{\xi}) = \sum_{i=1}^{d} |\xi_i|$, with the function to be minimized $\mathcal{L}(\mathbf{y}, \mathbf{y}') + \lambda \mathbb{P}(\boldsymbol{\xi})$.

4.2 Variants

Among the many variants, looking to enhance robustness with respect to noise, stability with respect to the slight variations in the inputs, variational, denoising and contractive autoencoders represent appealing choices, some of them summarized below.

4.2.1 Variational Autoencoders

In variational autoencoders, VAE, (an excellent step-by-step introduction can be found in [7]) the input data is encoded as distributions enabling continuity and completeness. In practice, normal distributions are chosen, such that the encoder is trained to return the mean and the covariance matrix, both depending on the input data, that allows a statistical sampling on which the decoder applies.

The loss function considered for training the network, is composed of two terms, the first (the standard one) ensuring the reconstruction performance, and the second one that regularizes the organization of the latent layer to ensure that the encoded distribution approaches a standard normal distribution. Covariance close to the identity prevents punctual distributions and a mean close to zero ensures proximity.

Again, the most natural way of expressing that regularization consists of using the Kullback-Leibler divergence between the returned distribution and a standard normal distribution.

Thus, the input data serves to extract, at the latent space level, the statistical descriptors (regularized by the standard normal distribution), whose sampling becomes the input of the decoding step. The decoding mean is learnt, while the covariance is given and represents the compromise between reproduction performances and regularization.

4.2.2 Denoising Autoencoders

In order to obtain representations able to filter noise, denoising AE, DAE, employ a training data set $\mathbf{y}_i$ which is previously corrupted from a stochastic map: $\mathbf{y} \rightarrow \mathbf{y}'$. Then, the AE rationale applies in the standard manner, while enforcing to recover the original (denoised) data $\mathbf{y}$.

4.2.3 Contractive Autoencoders

Contractive AE, CAE, look to improve robustness with respect to slight variations in the inputs, and for that purpose consider as penalty the Frobenius norm of the Jacobian, i.e. $\mathcal{L}(\mathbf{y}, \mathbf{y}') + \lambda \sum_{i=1}^{d} \|\nabla_y \xi_i\|^2$.

References

1. I. Goodfellow, Y. Bengio, A. Courville, *Deep Learning* (MIT Press, Cambridge, 2016)
2. J. Schmidhuber, Deep learning in neural networks: an overview. Neural Netw. **61**, 85–117 (2015)
3. G.E. Hinton, R.S. Zemel, Autoencoders, minimum description length and Helmholtz free energy, in *Advances in Neural Information Processing Systems*, vol. 6 (NISP 1993) (Morgan-Kaufmann, 1994), pp. 3–10
4. Q. Hernandeza, A. Badias, D. Gonzalez, F. Chinesta, E. Cueto, Deep learning of thermodynamics-aware reduced-order models from data. J. Comput. Phys. **426**, 109950 (2021)
5. D.P. Kingma, M. Welling, An introduction to variational autoencoders. Found. Trends Mach. Learn. **12**(4), 307–392 (2019)
6. A. Makhzani, B. Frey, k-sparse autoencoders (2014). arXiv:1312.5663v2
7. J. Rocca, Understanding variational autoencoders (VAEs). https://towardsdatascience.com/understanding-variational-autoencoders-vaes-f70510919f73

Open Access This chapter is licensed under the terms of the Creative Commons Attribution-NonCommercial-NoDerivatives 4.0 International License (http://creativecommons.org/licenses/by-nc-nd/4.0/), which permits any noncommercial use, sharing, distribution and reproduction in any medium or format, as long as you give appropriate credit to the original author(s) and the source, provide a link to the Creative Commons license and indicate if you modified the licensed material. You do not have permission under this license to share adapted material derived from this chapter or parts of it.

The images or other third party material in this chapter are included in the chapter's Creative Commons license, unless indicated otherwise in a credit line to the material. If material is not included in the chapter's Creative Commons license and your intended use is not permitted by statutory regulation or exceeds the permitted use, you will need to obtain permission directly from the copyright holder.

Chapter 5
Tensor Formats and Tensor Decompositions

Tensor decompositions allow us to express a high-order tensor as a product of lower dimensional tensors.

Singular Value Decomposition, SVD, is very well know for factorizing matrices. The extension of SVD to high-dimensional settings, which is referred to as high order singular value decomposition, HOSVD, proceeds with higher-dimensional tensors.

Proper Generalized Decomposition, PGD, that will be addressed in detail in the present monograph, can be viewed as a rank-one constructor, a particular case of more general decompositions (e.g. Tucker decomposition or the tensor trains) [1–7].

5.1 SVD and HOSVD Versus PGD

We begin by reviewing a classical technique, the Singular Value Decomposition (SVD). Let $\mathbf{M}$ be a $\mathcal{I} \times \mathcal{J}$ matrix, whose entries will be denoted by M_{ij}, that is assumed to be of rank r. The SVD factorizes $\mathbf{M}$ in the form $\mathbf{M} = \mathbf{U}\mathbf{\Sigma}\mathbf{V}^T$. The σ_i's (the components of the diagonal matrix $\mathbf{\Sigma}$) are the singular values of $\mathbf{M}$, ordered in descending order, such that σ_1 is the largest one.

SVD can thus then be seen as the decomposition of a matrix as a weighted, ordered sum of separate matrices

$$\mathbf{M} = \sum_{i=1}^{r} \mathbf{M}_i = \sum_{i=1}^{r} \sigma_i \mathbf{U}_i \mathbf{V}_i^T.$$

This separated representation can also be computed within the so-called PGD framework by minimizing $\|\mathbf{M} - \sum_i \mathbf{F}_i \mathbf{G}_i^T\|$ [8].

The solution procedure assumes at iteration n the $(n-1)$-rank approximation $\mathbf{M}^{n-1} = \sum_{i=1}^{n-1} \mathbf{F}_i \mathbf{G}_i^T$ already calculated, looking for the update $\mathbf{M}^n = \mathbf{M}^{n-1} + \mathbf{F}_n \mathbf{G}_n^T$, that entails a new iteration loop. In turn, $\mathbf{F}_n$ is calculated while assuming the $\mathbf{G}_n$ coming from the previous iteration of the nonlinear solution procedure, and then updating $\mathbf{G}_n$ from the just calculated $\mathbf{F}_n$, both calculations looking to minimize $\|\mathbf{M} - \mathbf{M}^n\|$.

In two dimensions, SVD and PGD have been demonstrated to be equivalent. Both provide an optimal separated representation of a given matrix. In three or more dimensions, however, both the high-order singular value decomposition, HOSVD, and the PGD produce compact separated representations of a given function, but their optimality is not guaranteed [8].

SVD (or PGD) can be visualized as follows: For a given matrix $\mathbf{M}$ we consider a column vector $\mathbf{C}_1$ and a row vector $\mathbf{R}_1$, offering the best approximation in the Frobenius norm, i.e.,

$$\{\mathbf{C}_1, \mathbf{R}_1\} = \arg\min_{\{\mathbf{C}, \mathbf{R}\}} \|\mathbf{M} - \mathbf{C}\mathbf{R}^T\|.$$

Then, the second pair of vectors are computed to best approximate the one-mode residual $\mathbb{R}_1 = \mathbf{M} - \mathbf{C}_1 \mathbf{R}_1^T$, i.e.

$$\{\mathbf{C}_2, \mathbf{R}_2\} = \arg\min_{\{\mathbf{C}, \mathbf{R}\}} \|\mathbb{R}_1 - \mathbf{C}\mathbf{R}^T\|,$$

and so on.

5.2 CUR Decomposition

To avoid the complexity of solving a eigenproblem (SVD) or the just described optimization problems (PGD), the so-called CUR approximation [9] proposes to select a number of columns N_c of $\mathbf{M}$, organized in the columns of matrix $\mathbb{C}$, and a number of rows N_r of $\mathbf{M}$, disposed in the rows of matrix $\mathbb{R}$. Then, the CUR looks for the matrix $\mathbb{U}$, of size $N_r \times N_c$, minimizing

$$\mathbb{U} = \arg\min_{\hat{\mathbb{U}}} \|\mathbf{M} - \mathbb{C}\hat{\mathbb{U}}\mathbb{R}\|_2.$$

Even if in general the rows and columns are randomly chosen, more elaborated procedures to choose them exist.

5.3 Cross Approximation

Cross approximation can be viewed as a strategy to drive the columns and rows selection within the rationale descried in the previous section [10, 11].

We first identify the column c_1 and row r_1 associated to the maximum component of matrix $\mathbf{M}$, i.e. $|M_{r_1,c_1}| \geq |M_{i,j}|$, $\forall i, j$. The first selected column-row pair consists of $\mathbf{C}_1 = \mathbf{M}_{:,c_1}$ and $\mathbf{R}_1 = \mathbf{M}_{r_1,:}$.

Thus, the first residual reads $\mathbb{R}_1 = \mathbf{M} - d_1^1 \mathbf{C}_1 \mathbf{R}_1^T$, where coefficient d_1^1 is calculated in order to ensure that the (r_1, c_1) component of $\mathbb{R}_1$, $\mathbb{R}_{1_{r_1,c_1}}$, vanishes.

Now, the maximum of $\mathbb{R}_1$ determines the second column-row pair (c_2, r_2), that necessarily differs from the first due to the fact that $\mathbb{R}_{1_{r_1,c_1}} = 0$. Then, the new residual reads $\mathbb{R}_2 = \mathbf{M} - d_1^2 \mathbf{C}_1 \mathbf{R}_1^T - d_2^2 \mathbf{C}_2 \mathbf{R}_2^T$, where coefficients d_1^2 and d_2^2 are calculated to ensure that $\mathbb{R}_2$ vanishes at entrees (r_1, c_1) and (r_2, c_2).

The process continues until the maximum of the residual at iteration n (after n rows and columns have been introduced) becomes small enough, with $\mathbf{M}$ approximated by

$$\mathbf{M} = d_1^n \mathbf{C}_1 \mathbf{R}_1^T + \cdots + d_n^n \mathbf{C}_n \mathbf{R}_n^T,$$

and the coefficients $d_1^n, \ldots, d_n^n$ ensuring that $\mathbb{R}_n$ vanishes at entrees $(r_1, c_1), \ldots, (r_n, c_n)$.

References

1. C.F. Van Loan, The ubiquitous Kronecker product. J. Comput. Appl. Math. **123**, 85–100 (2000)
2. T.G. Kolda, B.W. Bader, Tensor decompositions and applications. SANDIA Report, SAND2007-6702 (2007)
3. I.V. Oseledets, Tensor-train decomposition. SIAM J. Sci. Comput. **33**(5), 2295–2317 (2011)
4. W. Hackbusch, *Tensor Spaces and Numerical Tensor Calculus* (Springer, Berlin Heidelberg, 2012)
5. L. Grasedyck, D. Kressner, C. Tobler, A literature survey of low-rank tensor approximation techniques. https://arxiv.org/abs/1302.7121
6. B.N. Khoromskij, Tensor numerical methods for high-dimensional PDEs: basic theory and initial applications. https://doi.org/10.48550/arXiv.1408.4053
7. N.D. Sidiropoulos, L. De Lathauwer, X. Fu, K. Huang, E.E. Papalexakis, C. Faloutsos, Tensor decomposition for signal processing and machine learning. IEEE Trans. Signal Process. **65**(13), 3551–3582 (2017)
8. F. Chinesta, R. Keunings, A. Leygue, *The Proper Generalized Decomposition for Advanced Numerical Simulations: A Primer* (Springer, . Springerbriefs, 2014)
9. M.W. Mahoney, P. Drineas, CUR matrix decompositions for improved data analysis. PNAS **106**(3), 697–702 (2009)
10. M. Espig, L. Grasedyck, W. Hackbusch, Black box low tensor-rank approximation using fiber-crosses. Constr. Approx. **30**, 557–597 (2009)
11. J.V. Aguado, D. Borzacchiello, K.S. Kollepara, F. Chinesta, A. Huerta, Tensor representation of on-linear models using cross approximations. J. Sci. Comput. **81**, 22–47 (2019)

Open Access This chapter is licensed under the terms of the Creative Commons Attribution-NonCommercial-NoDerivatives 4.0 International License (http://creativecommons.org/licenses/by-nc-nd/4.0/), which permits any noncommercial use, sharing, distribution and reproduction in any medium or format, as long as you give appropriate credit to the original author(s) and the source, provide a link to the Creative Commons license and indicate if you modified the licensed material. You do not have permission under this license to share adapted material derived from this chapter or parts of it.

The images or other third party material in this chapter are included in the chapter's Creative Commons license, unless indicated otherwise in a credit line to the material. If material is not included in the chapter's Creative Commons license and your intended use is not permitted by statutory regulation or exceeds the permitted use, you will need to obtain permission directly from the copyright holder.

Chapter 6
Dictionary Learning

Dictionary Learning, DL, tries to express a series of data as a sparse combination of the elements stored in a dictionary. The dictionary is noted by $\mathbf{D} = [\mathbf{d}_1, \ldots \mathbf{d}_M] \in \mathbb{R}^{d \times M}$, the data $\mathbf{x}_i \in \mathbb{R}^d$, $i = 1, \ldots, K$ are grouped in the columns of matrix $\mathbf{X}$, and the data representation in the dictionary by vectors $\mathbf{r}_j \in \mathbb{R}^M$, $j = 1, \ldots, K$.

Given the input data $\mathbf{X}$, DL finds the dictionary as well as the data representation by minimizing the representation error $\|\mathbf{X} - \mathbf{DR}\|_2^2$ (with the Frobenius norm), while keeping the representation sparse enough, i.e.

$$\{\mathbf{D}, \mathbf{R}\} = \arg\min_{\hat{\mathbf{D}} \in \mathcal{C},\ \hat{\mathbf{r}}_i \in \mathbb{R}^M} \sum_{i=1}^{K} \|\mathbf{x}_i - \hat{\mathbf{D}}\hat{\mathbf{r}}_i\|_2^2 + \lambda \|\hat{\mathbf{r}}_i\|_0,$$

with $\mathcal{C}$ the matrices with $\|\mathbf{d}_i\|_2 \leq 1$, $i = 1, \ldots, M$, condition that limits an increase of the dictionary components induced by the enforced sparsity of the representation vectors.

However, it is well known that enforcing a L0-norm leads to a non-convex minimization problem. Thus, in general it is replaced by a L1-norm that enforces sparsity (as we will discuss later when addressing compressed sensing in Sect. 10).

In particular DL considers different techniques to solve the minimization problem with the L0-norm sparsity constraint. Among them [1]:

- The *Method of Optimal Directions*, that alternates between computing the sparse representation (using the L0-norm) using for example the *Matching Pursuit* algorithm, and then computing the dictionary expression from $\mathbf{D} = \mathbf{XR}^+$, with $\mathbf{R}^+$ the Moore-Penrose pseudo inverse;
- The kSVD algorithm, that first finds the best sparse representation again considering the L0-norm (e.g. by using the Matching Pursuit) and then iteratively updating one element of the dictionary by computing the rank-1 approximation of the residual when the element that is being updated was removed from the dictionary, while ensuring sparsity;

- The use of the *Stochastic Gradient Descent*, that updates the dictionary according to

$$\mathbf{D}_i = \text{Proj}_\mathcal{C} \left\{ \mathbf{D}_{i-1} - \delta_i \nabla_\mathbf{D} \sum_{j \in \mathcal{S}} \|\mathbf{x}_j - \mathbf{D}\mathbf{r}_j\|_2^2 \right\},$$

where $\mathcal{S}$ is a random subset of $\{1, \ldots, K\}$ and δ_i is a gradient step. The previous gradient is usually regularized. From the updated dictionary, the sparse representation is calculated. The iteration continues until reaching convergence.

Reference

1. A. Moitra, *Algorithmic aspects of machine learning* (Cambridge University Press, 2018)

Open Access This chapter is licensed under the terms of the Creative Commons Attribution-NonCommercial-NoDerivatives 4.0 International License (http://creativecommons.org/licenses/by-nc-nd/4.0/), which permits any noncommercial use, sharing, distribution and reproduction in any medium or format, as long as you give appropriate credit to the original author(s) and the source, provide a link to the Creative Commons license and indicate if you modified the licensed material. You do not have permission under this license to share adapted material derived from this chapter or parts of it.

The images or other third party material in this chapter are included in the chapter's Creative Commons license, unless indicated otherwise in a credit line to the material. If material is not included in the chapter's Creative Commons license and your intended use is not permitted by statutory regulation or exceeds the permitted use, you will need to obtain permission directly from the copyright holder.

Chapter 7
Time Series: Metrics and Alignment

7.1 Standard Metrics for Time Series

Most standard procedures and metrics for manipulating time series were reviewed in [1] and are summarized in the present section.

Denoting by D the time series similarity measure, its calculation can be grouped in three main categories depending on the way it performs: (i) operating directly with the original time series data; (ii) operating on some transformation performed on the original time series; and (iii) operating indirectly from metrics derived from the original data.

1. *Approaches applying directly on the time series.*

   The most usual metrics concern the use of distances or correlations. In the former, the Minkowski distance $D_{\texttt{Mink}}$ is widely considered. When applied on two data series $\mathbf{f}^p$ and $\mathbf{f}^q$, with components f_n^p and f_n^q, with $f_n \equiv f^{\cdot}(t = n\Delta t)$, it reads

$$D_{\texttt{Mink}} = \left(\sum_{n=0}^{N-1} |f_n^p - f_n^q|^r \right)^{1/r},$$

   that for $r = 1$ and $r = 2$ leads to the Manhattan $D_{\texttt{Man}}$ and the Euclidean $D_{\texttt{E}}$ distances respectively, the last, more sensitive to outliers.

   To reduce the effect of the observations with largest variance, the Mahalanobis distance $D_{\texttt{Mah}}$ is usually considered:

$$D_{\texttt{Mah}} = \sqrt{(\mathbf{f}^p - \mathbf{f}^q)^T \Sigma^{-1} (\mathbf{f}^p - \mathbf{f}^q)},$$

   where the covariance matrix Σ must be estimated beforehand.

   Concerning the correlations, the Pearson cross correlation $D_{\texttt{CC}}$ is one of the most widely considered. It reads:

$$D_{\text{CC}}(s) = \frac{\sum_{n=0}^{N-1}(f_n^p - \overline{f^p}) \cdot (f_{n-s}^q - \overline{f^q})}{\sqrt{\sum_{n=0}^{N-1}(f_n^p - \overline{f^p})^2} \cdot \sqrt{\sum_{n=0}^{N-1}(f_{n-s}^q - \overline{f^q})^2}},$$

where the integer s, $s \geq 0$, reflects the considered shift.

2. *Approaches based on the time series transformation.*
   A first approach consists in using the PCA, to extract the most dominant modes, and then project each time series on those modes to obtain the m largest principal components $PC_k^{\cdot}$, $k = 1, \ldots, m$; that allows defining the so-called PCA-based metrics D_{PCA}

$$D_{\text{PCA}} = \sqrt{\sum_{k=1}^{m}(PC_k^p - PC_k^q)^2}.$$

Another possibility consists in applying a Fourier transform to each time series and then, computing the Fourier distance D_{Fou} from the Fourier coefficients.

3. *Approaches based on derived metrics.*
   These approaches convert the raw time-data series into a number of indicators that describe the time series by using simple statistics, and subsequently calculate similarity measures indirectly through these statistics. These descriptors are goal oriented and depend strongly on the physics behind the analyzed process. Among the abundant choices, local average, standard deviation, pair correlation, covariogram, among others, are widely considered.

7.2 Dynamic Time Warping

Dynamic Time Warping, DTW, tries to minimize the effects of shifting and distortion in time series [2, 3]. We begin by considering two data sets, $\mathcal{X} = \{x_1, x_2, \ldots, x_N\}$, $N \in \mathbb{N}$ and $\mathcal{Y} = \{y_1, y_2, \ldots, y_M\}$, $M \in \mathbb{N}$. We first compute the distance matrix $\mathbf{D}$, also called cost matrix, with components $D_{ij} = |x_i - y_j|$, $i = 1, \ldots, N$ and $j = 1, \ldots M$.

The so-called *alignment path* identifies the low-cost areas or, in other words, the *valleys* on the cost matrix. Mathematically, the path is composed by the sequence of points $\mathcal{P} = \{\mathbf{p}_1, \mathbf{p}_2, \ldots, \mathbf{p}_K\}$ with $\mathbf{p}_l = (x_{m_l}, y_{n_l})$, $x_{m_l} \in \mathcal{X}$, $y_{n_l} \in \mathcal{Y}$ and $l = 1, \ldots, K$, satisfying:

- $\mathbf{p}_1 = (x_{m_1}, y_{n_1}) = (x_1, y_1)$ and $\mathbf{p}_K = (x_{m_K}, y_{n_K}) = (x_N, y_M)$;
- Monotonicity implies that in the series $\mathcal{P}$, $x_{m_{l-1}} < x_{m_l} < x_{m_{l+1}}$ and $y_{n_{l-1}} < y_{n_l} < y_{n_{l+1}}$;
- Step condition: $(m_l, n_l) - (m_{l-1}, n_{l-1}) \in \{(1, 1), (1, 0), (0, 1)\}$.

The path cost (to be minimized) reads

$$\mathcal{C} = \sum_{l=1}^{K} D_{m_l,n_l}.$$

The alignment path is calculated by dynamic programming. For that purpose we define the matrix of accumulated distances $\mathbf{A}$:

- First row:

$$A_{1,j} = \sum_{k=1}^{j} D_{1,k};$$

- First column:

$$A_{i,1} = \sum_{k=1}^{i} D_{k,1};$$

- The remaining components:

$$A_{i,j} = \min\{A_{i-1,j-1}, A_{i-1,j}, A_{i,j-1}\} + D_{i,j};$$

Now, starting from the terminal point, the path is reconstructed backwards.

7.3 Optimal Transport and Topological Data Analysis

Optimal transport, OT, also serves to compare data series and Topological Data Analysis, TDA, will be addressed in a later stage.

References

1. S. Lhermitte, J. Verbesselt, W.W. Verstraeten, P. Coppin, A comparison of time series similarity measures for classification and change detection of ecosystem dynamics. Remote Sens. Environ. **115**, 3129–3152 (2011)
2. M. Muller, *Information retrieval for music and motion* (Springer, Berlin, Heidelberg, 2007)
3. P. Senin, Dynamic time warping algorithm review. Technical report (2008)

Open Access This chapter is licensed under the terms of the Creative Commons Attribution-NonCommercial-NoDerivatives 4.0 International License (http://creativecommons.org/licenses/by-nc-nd/4.0/), which permits any noncommercial use, sharing, distribution and reproduction in any medium or format, as long as you give appropriate credit to the original author(s) and the source, provide a link to the Creative Commons license and indicate if you modified the licensed material. You do not have permission under this license to share adapted material derived from this chapter or parts of it.

The images or other third party material in this chapter are included in the chapter's Creative Commons license, unless indicated otherwise in a credit line to the material. If material is not included in the chapter's Creative Commons license and your intended use is not permitted by statutory regulation or exceeds the permitted use, you will need to obtain permission directly from the copyright holder.

Chapter 8
Data Separation: Independent Component Analysis

We consider multi-dimensional data, representing a linear mixture of the hidden sources **s**. For example each component of **s**, s_j, could represent a sound source at a certain time (music, voice, noise, ...) and each component of the data **x**, x_k represents the sound reception at different spatial location, at the corresponding time. Thus, each receptor mixes the different emission sources. A different vector $\mathbf{x}_i$, can be collected at each time instant t_i, from the associated sources $\mathbf{s}_i$.

We assume M different data vectors $\mathbf{x}_i$, related to the same number of hidden vectors $\mathbf{s}_i$, according to the mixing matrix **A**, such that $\mathbf{X} = \mathbf{AS}$, or $\mathbf{S} = \mathbf{A}^{-1}\mathbf{X}$, where the columns of matrix **X** and **S** contain vectors $\mathbf{x}_i$ and $\mathbf{s}_i$.

Since the sum of two independent random variables is more Gaussian than the individual original random variables, maximizing non-Gaussianity will be one of the key ingredients of Independent Component Analysis, ICA [1–3]. Among the techniques that perform in that sense, statistical kurtosis is one option (even if it is quite sensitive to outliers).

Some data pre-processing must preceed the calculation of both **A** and **S**:

- Data-centering: $\mathbf{X} \leftarrow \mathbf{X} - \mathbb{E}\{\mathbf{X}\}$, where without risk of confusion the normalized data is still denoted by **X**;
- Whitening (also called *sphearing*): We decompose the covariance matrix $\mathbb{E}(\mathbf{XX}^T) = \mathbf{UDU}^T$, and then transform the centered data according to:

$$\mathbf{X} \leftarrow (\mathbf{D}^{-1/2}\mathbf{U}^T)\mathbf{X},$$

where again the whitened data is still denoted by **X**.
Whitening ensures $\mathbb{E}(\mathbf{XX}^T) = \mathbf{I}$. To prove it, it suffices writing $(\mathbf{D}^{-1/2}\mathbf{U}^T)(\mathbf{XX}^T)\mathbf{UD}^{-1/2}$, applying $\mathbb{E}(\cdot)$ and taking into account that (with the data before whitening) $\mathbb{E}(\mathbf{XX}^T) = \mathbf{UDU}^T$.

With matrix $\mathbf{A}^{-1}$ expressed as $\mathbf{VD}^{-1/2}\mathbf{U}^T$, matrix $\mathbf{V}$, verifying the orthonormality condition $\mathbf{V}^T\mathbf{V} = \mathbf{I}$, is searched to maximize the non-gaussianity of $\mathbf{VX}$, with $\mathbf{X}$ centered and whitened.

References

1. A. Hyvarinen, E. Oja, Independent component analysis: algorithms and applications. Neural Netw. **13**(4–5), 411–430 (2000)
2. N. Kutz, *Data-driven modeling and scientific computation: methods for complex systems and big data* (Oxford University Press, Oxford, 2013)
3. J. Shlens, *A tutorial on independent component analysis*. arXiv:1404.2986v1

Open Access This chapter is licensed under the terms of the Creative Commons Attribution-NonCommercial-NoDerivatives 4.0 International License (http://creativecommons.org/licenses/by-nc-nd/4.0/), which permits any noncommercial use, sharing, distribution and reproduction in any medium or format, as long as you give appropriate credit to the original author(s) and the source, provide a link to the Creative Commons license and indicate if you modified the licensed material. You do not have permission under this license to share adapted material derived from this chapter or parts of it.

The images or other third party material in this chapter are included in the chapter's Creative Commons license, unless indicated otherwise in a credit line to the material. If material is not included in the chapter's Creative Commons license and your intended use is not permitted by statutory regulation or exceeds the permitted use, you will need to obtain permission directly from the copyright holder.

Chapter 9
Topological Data Analysis

In this section we revisit the topological descriptors of data series and images. These descriptors consist of an alternative or complementary route to more experienced statistical descriptors (moments, correlations, covariograms, ...) [1]. They enable comparing, clustering, classifying and modeling (from nonlinear regressions) data with high topology content.

These topological descriptors enable their manipulation in appropriate vector spaces, making possible evaluating proximity, or facilitating the construction of models where images or time-series represent the model inputs and/or model outputs. This section reviews Topological Data Analysis, TDA [2–6].

9.1 Illustrating TDA on a Time Series

Topological Data Analysis, TDA, applied to time series transforms the data-point series into a compact representation in a vector space. To illustrate the method, we consider a simple example, by considering a data set consisting in nine integers,

$$S = \{11, 14, 9, 7, 9, 7, 8, 10, 9\},$$

corresponding, respectively, to the points in the plane

$$\{(0, 11), (1, 14), (2, 9), (3, 7), (4, 9), (5, 7), (6, 8), (7, 10), (8, 9)\}.$$

The time series contains 4 local minima: $\{(0, 11), (3, 7), (5, 7), (8, 9)\}$; and 3 local maxima: $(1, 14), (4, 9), (7, 10)$.

Now, minimum-maximum neighbors are paired: $(0, 11)$ to $(1, 14)$, then $(8, 9)$ to $(7, 10)$ and finally $(5, 7)$ to $(4, 9)$.

The local minimum-local maximum paired values gives rise to the so-called persistence diagram $\mathcal{PD}(S)$. In this toy example, $\mathcal{PD}(S) = \{(7, 9), (9, 10), (11, 14)\}$. The minimum of each pair represents the topological occurrence birth, while the associated maximum represents its death.

The persistence diagram is thus a two-dimensional representation in which each data-point (x, y) satisfies the obvious relationship $y \geq x$ (the topological occurrence birth precedes its death). Consequently, every point is located above the bisector $x = y$. The persistence diagram offers, therefore, a valuable and concise description of any curve (e.g. surface profiles, time-series, ...) and the change in their topological features.

In a strictly related setting, the *lifetime diagram* $\mathcal{T}(S)$ associated to every $\mathcal{PD}(S)$, is defined as $\mathcal{T}(S) = \{(x, y) \in \mathcal{PD}(S) \to (x, y - x) \in \mathbb{R}^2\}$, where $y - x$ represents the lifetime of the topological occurrence. In the toy problem introduced before, it turns out that $\mathcal{T}(S) = \{(7, 2), (9, 1), (11, 3)\}$.

Next, we can derive the persistence image $\mathcal{PI}(S)$, that is defined in a vector space. For that purpose a continuous piecewise differentiable non-negative weight function $w(x, y)$ is considered, with $(x, y) \in \mathcal{T}(S)$, $w(x, 0) = 0$ and $w(x, y_{max}) = 1$, with $y_{max} = \max(y)$, approximated from a linear function of the lifetime y, e.g. $w(x, y) = y/y_{max}$, and a bivariate normal distribution $g_{x,y}(u, v)$ centered at point $(x, y) \in \mathcal{T}(S)$ and with a variance σ, $\sigma > 0$, that scales with the maximum value of the lifetime diagram.

We define

$$\rho_S(u, v) = \sum_{(x,y) \in \mathcal{T}(S)} w(x, y)\, g_{(x,y)}(u, v),$$

with $(u, v) \in \mathcal{D}$, with $\mathcal{D}$ a compact domain, typically, the domain of definition of $\mathcal{T}(S)$.

We proceed now by partitioning the domain $\mathcal{D}$ into a series of non-overlapping covering subdomains, often referred to as pixels P_i, with $\mathcal{D} = \cup_{i=1}^{P} P_i$. A function $\rho_S(u, v)$ is then averaged at each of these pixels to define the *persistence image* $\mathcal{PI}(S)$. In this way, each P pixel of the persistence image $\mathcal{PI}(S)$ takes the value

$$\mathcal{PI}_{P_i}(S) = \iint_{P_i} \rho_S(u, v)\, du\, dv.$$

The persistence image described the time series while heritages all the invariance properties of topology. Thus, it consists in an appealing description to be considered in data-analytics and machine learning.

9.2 Filtration and Homology Group

The TDA construction can be formalized from the use of the so-called filtrations, on which the homology group applies.

9.2.1 Rips Filtration

We consider a set $\mathbb{S}$, composed of points x_i, for simplicity in $\mathbb{R}^2$. We construct a *Rips Complex* by grouping simplices of different dimensions. A d-simplex is the smallest convex set of $d+1$ points, $x_0, \ldots, x_d$ where $x_1 - x_0, \ldots, x_d - x_0$ are linearly independent, see Fig. 9.1.

The Rips complex $\mathcal{R}_\epsilon(\mathbb{S})$ can be constructed by considering balls of radius $\frac{\epsilon}{2}$, centered at each point as illustrated in Fig. 9.2.

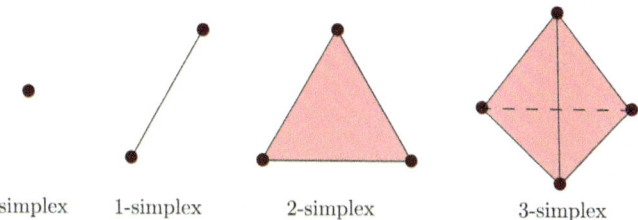

Fig. 9.1 Simplices of different dimensions

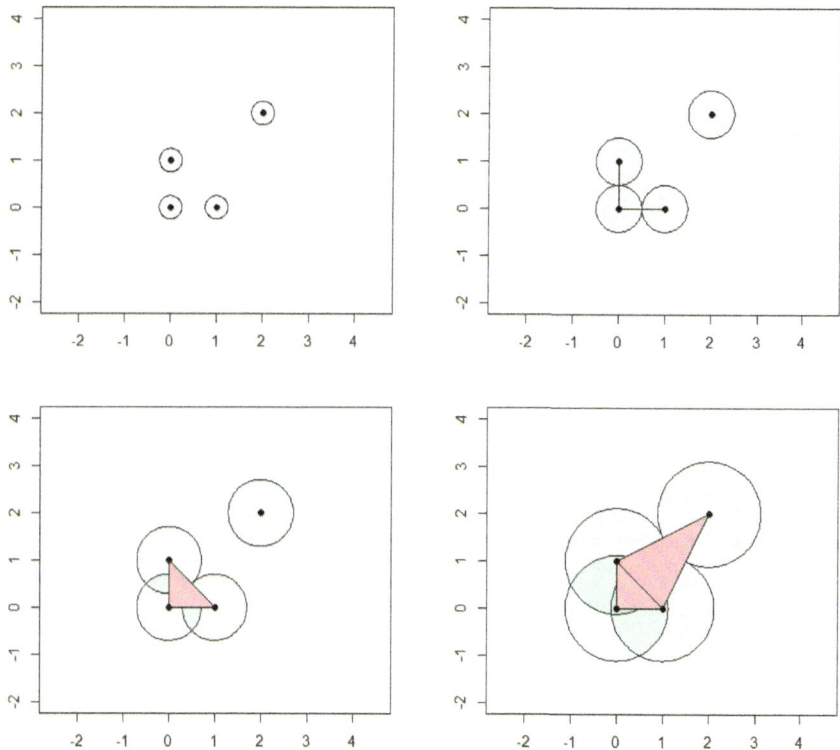

Fig. 9.2 Example of Rips complex computation: (top-left) $\epsilon = 0.5$; (top-right) $\epsilon = 1$; (bottom-left) $\epsilon = 1.4$; and (bottom-right) $\epsilon = 2.3$

Consider now a particular Homology group. To track the appearance of the features across different scales, we define the Homology group at a scale ϵ, $H_k^\epsilon(\mathbb{S})$. This homology group will represent the classes of simplices, but this time taken from $\mathcal{R}_\epsilon(\mathbb{S})$. In other words, it represents the elements of $\mathcal{R}_\epsilon(\mathbb{S})$ with a filtration value lower than ϵ. Persistent homology allows quantifying the appearance and disappearance of the features across the different scales, discretized by considering m values related to ϵ_j, $j = 0, \ldots, m$:

- For $H_0(\mathbb{S})$, the birth scale of all vertices is set to zero, while the death scale is the filtration value at which the vertex has been joined to another one by a segment.
- For $H_1(\mathbb{S})$, the birth scale of a cycle is the filtration value at which a loop has been formed, while the death scale is the filtration value at which the interior of the loop has been covered.

In other words,

- The birth scale b_γ of the feature γ

$$b_\gamma = \min_{0 \leq j \leq m} \{\epsilon_j : \gamma \in H_k^{\epsilon_j}\}$$

- The death scale d_γ of the feature γ

$$d_\gamma = \max_{0 \leq j \leq m} \{\epsilon_j : \gamma \in H_k^{\epsilon_j}\}$$

The persistence of each feature through the different scales can alternatively be represented by the so-called persistence barcode of $\mathbb{S}$. This barcode constitutes a histogram, with the bar associated to each feature starting at the birth scale and ending at the death scale.

An example of persistent homology computation is given with the Rips complex shown in Fig. 9.3, and its associated persistence diagram

$$\mathcal{PD}(\mathbb{S}) = \{(b_\gamma, d_\gamma) : \gamma \in H_k\},$$

depicted in Fig. 9.4, where b_γ and d_γ represent the birth and death scales associated to the feature γ, respectively. It can be noticed that a loop is formed at $\epsilon = 0.9$ (birth) and then covered at $\epsilon = 1.8$ (death). It is indicated by the red bar.

This filtration was successfully applied to analyze complex microstructures and performing subsequent regressions in [7] and for extracting trajectory patterns for diagnosis purposes in [8]. Rips filtration also allowed filtering noise in data prior to perform machine learning [9].

9.2.2 Other Filtrations

Sub-Level-Set-based filtration, closely related to the Reeb graph [10] used in Morse theory, was used in [11], whereas *alpha filtration* was employed in [12] to efficiently

9.2 Filtration and Homology Group

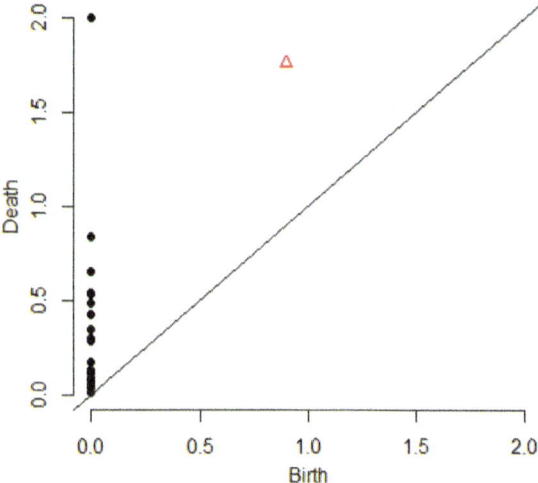

Fig. 9.3 Example of Rips complex computation: (top-left) $\epsilon = 0$; (top-right) $\epsilon = 0.5$; (bottom-left) $\epsilon = 0.9$; and (bottom-right) $\epsilon = 1.8$

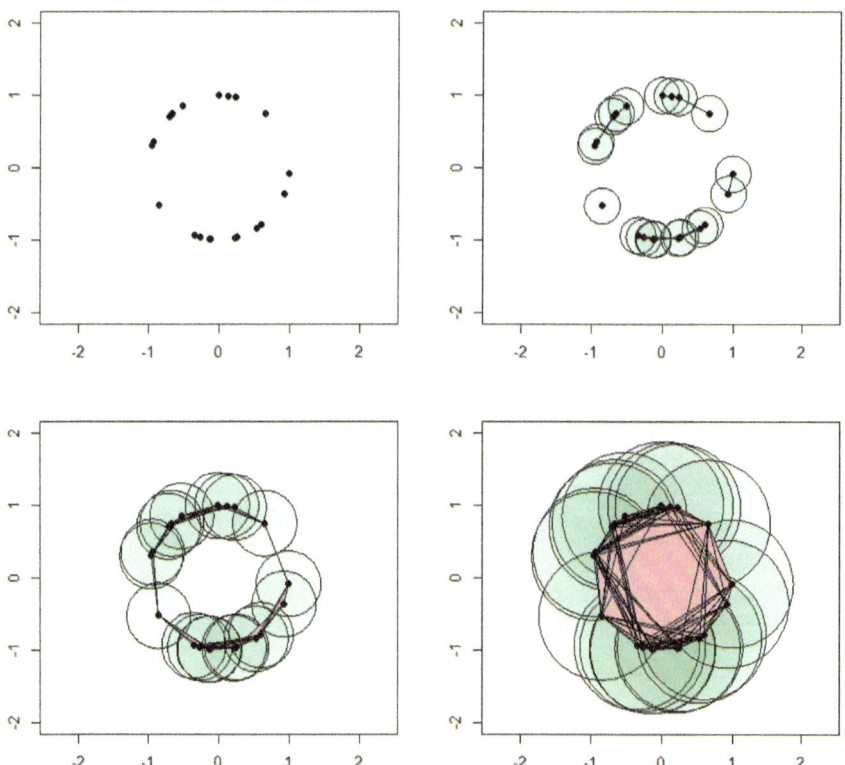

Fig. 9.4 Persistence Diagram: in black the H_0 features, and in red the H_1 feature

describe the geometry of the dataset related to the points on a surface embedded into $\mathbb{R}^3$, for classifying structural dynamics eigenmodes.

References

1. S. Torquato, Statistical description of microstructures. Annu. Rev. Mater. Res. **32**, 77–11 (2002)
2. G. Carlsson, A. Zomorodian, A. Colling, L. Guibas, Persistence barcodes for shapes. *Eurographics Symposium on Geometry Processing* (2004)
3. G. Carlsson, Topology and data. Bull. Am. Math. Soc **46**(2), 255–308 (2009)
4. F. Chazal, B. Michel, An introduction to Topological Data Analysis: fundamental and practical aspects for data scientists. J de la Societe Francaise de Statistiques (2017)
5. L. Wasserman, Topological data analysis. Ann. Rev. Stat. Appl. **5**(1), 501–532 (2018)
6. N. Saul, C. Tralie, *Scikit-TDA: Topological Data Analysis for Python* (2019)
7. M. Yun, C. Argerich, E. Cueto, J.L. Duval, F. Chinesta, Nonlinear regression operating on microstructures described from Topological Data Analysis for the real-time prediction of effective properties. Materials **13**(10), 2335 (2020)
8. T. Frahi, A. Sancarlos, M. Galle, X. Beaulieu, A. Chambard, A. Falco, E. Cueto, F. Chinesta. Monitoring weeder robots and anticipating their functioning by using advanced topological data analysis. Front Artif Intell **4**
9. B. Moya, D. Gonzalez, I. Alfaro, F. Chinesta, E. Cueto, Learning slosh dynamics by means of data. Comput. Mech. **64**, 511–523 (2019)
10. G. Reeb, Sur les points singuliers d'une forme de Pfaff complètement intégrable ou d'une fonction numérique. C. R. Acad. Sci. **222**, 847–849 (1946)
11. T. Frahi, F. Chinesta, A. Falco, A. Badias, E. Cueto, H.Y. Choi, M. Han, J.L. Duval, Empowering Advanced Driver-Assistance Systems from Topological Data Analysis. Mathematics **9**, 634 (2021)
12. T. Frahi, A. Falco, B. Vinh Mau, J.L. Duval, F. Chinesta, Empowering advanced parametric modes clustering from topological data analysis. Appl. Sci. **11**, 6554 (2021)

Open Access This chapter is licensed under the terms of the Creative Commons Attribution-NonCommercial-NoDerivatives 4.0 International License (http://creativecommons.org/licenses/by-nc-nd/4.0/), which permits any noncommercial use, sharing, distribution and reproduction in any medium or format, as long as you give appropriate credit to the original author(s) and the source, provide a link to the Creative Commons license and indicate if you modified the licensed material. You do not have permission under this license to share adapted material derived from this chapter or parts of it.

The images or other third party material in this chapter are included in the chapter's Creative Commons license, unless indicated otherwise in a credit line to the material. If material is not included in the chapter's Creative Commons license and your intended use is not permitted by statutory regulation or exceeds the permitted use, you will need to obtain permission directly from the copyright holder.

Chapter 10
Compressed Sensing

10.1 Sparsity

Very often, manifold learning techniques employ some form of local least-squares fitting of the data. The novelty in compressed sensing techniques relies in the use of the L1-norm instead [1]. The L1 norm constitutes an elegant way of enforcing sparsity. In regression, L2 norm gives an utmost importance to the outliers. This is due to the use of the squared norm, such that these outliers have a tremendous influence in the resulting fitted curve.

The issue is very similar to the solution of underdetermined algebraic systems of equations. Since they have an infinite number of solutions, the use of the pseudo-inverse produces a fully populated solution vector. The Matlab backslash, on the contrary, produces solutions that are sparser. Solving the problem by using L2- or L1-weighted optimizations produces, in the former case, a much less sparse solution than the last.

In some sense, L1-norm can be associated to sparsity. The L1-norm has been considered as an appealing technique to solve signal reconstruction problems. Its ability to overcome the Nyquist-Shannon sampling theorem (stating that in order to recover a signal, one must sample at twice the rate of the highest frequency involved in the signal) has been proved.

10.2 Sparse Sensing

Let $\mathbf{f}$ be a vector, in the space or time domains, and $\mathbf{c}$ its vector counterpart in a domain in which it should accept a sparse representation. These spaces are in general related to frequency (Fourier or discrete cosine transforms) or to multi-resolution wavelets, among others. Denote by $\mathbf{T}$ the matrix making possible this discrete transformation, so that $\mathbf{Tc} = \mathbf{f}$.

Given the fact that **c** is expected to have many zero entries (assuming that we are able to find the space in which the signal becomes sparse), we suposse that it could be found by employing only a few rows of matrix **T** and vector **f**. The resulting underdetermined system will therefore be solved by employing an L1-norm based optimization.

How choosing, however, the relevant rows for this purpose? The most usual algorithm employs a random selection. In matrix form, it is enough to define a diagonal matrix with unit entries at the rows we want to extract. Denoting the set of rows to be extracted by S, the extraction matrix **E** is defined from $E_{ii} = 1$ if $i \in S$ and $E_{ij} = 0$ otherwise. Rows containing only zeros are then eliminated from the matrix, producing a rectangular one, which we denote by **E**.

Solving the problem **Tc** = **f** can thus be done by solving the alternative underdetermined system **ETc** = **Ef**, by using an L1-norm based optimization.

Sparse sensing employs therefore two main ingredients: (i) an adequate space in which the solution of the problem is expected to admit a sparse representation, and (ii) the solution of the underdetermined problem by using the L1-norm.

Pushing compressed sensing to its limit we arrive at the so-called *single pixel camera* [2]. The usual camera procedure consists in acquiring a pixel vector **f**, that very often is then compressed, so that a few of its entries are registered, **Ef**. If we compute vector **c** by solving **ETc** = **Ef**, the whole image can be reconstructed from **f** = **Tc**. Similar approaches have been developed for model order reduction, for instance, see [3].

An alternative, and very useful way to solve **ETc** = **Ef** employs the so-called *least absolute shrinkage and selection operator*, LASSO, method [4]. LASSO involves an L^2- minimization of the system of equations plus a penalty term involving a L1-norm of the unknown field:

$$\mathbf{c} = \text{argmin}_{\mathbf{c}^*}(||\mathbf{ETc}^* - \mathbf{Ef}||_2^2 + \lambda ||\mathbf{c}^*||_1).$$

References

1. N. Kutz, *Data-driven modeling and scientific computation: methods for complex systems & big data* (Oxford University Press, Oxford, 2013)
2. M.F. Duarte, M.A. Davenport, D. Takhar, J.N. Laska, T. Sun, K.F. Kelly, R.G. Baraniuk, Single-Pixel imaging via compressive sampling. IEEE Signal Proc. Mag. **83** (2008)
3. R. Ibanez, E. Abisset-Chavanne, E. Cueto, A. Ammar, J.L. Duval, F. Chinesta, Some applications of compressed sensing in computational mechanics. Model order reduction, manifold learning, data-driven applications and nonlinear dimensionality reduction. Comput. Mech. **64**, 1259–1271 (2019)
4. R. Tibshirani, Regression shrinkage and selection via the lasso. J. R. Stat. Soc. Ser. B (Methodological) **58**(1), 267–288 (1996)

Open Access This chapter is licensed under the terms of the Creative Commons Attribution-NonCommercial-NoDerivatives 4.0 International License (http://creativecommons.org/licenses/by-nc-nd/4.0/), which permits any noncommercial use, sharing, distribution and reproduction in any medium or format, as long as you give appropriate credit to the original author(s) and the source, provide a link to the Creative Commons license and indicate if you modified the licensed material. You do not have permission under this license to share adapted material derived from this chapter or parts of it.

The images or other third party material in this chapter are included in the chapter's Creative Commons license, unless indicated otherwise in a credit line to the material. If material is not included in the chapter's Creative Commons license and your intended use is not permitted by statutory regulation or exceeds the permitted use, you will need to obtain permission directly from the copyright holder.

Chapter 11
Some Issues in Data Interpolation

This section revisits two tricky issues related to data interpolation. The first concerns complex data, where interpolating amplitude and phase angle is sometimes more robust than interpolating the data components independently. Second, we will address the issue related to data on a nonlinear manifold, where interpolation must be performed carefully.

11.1 Complex-Valued Variables

In problems involving electric and magnetic fields in electromagnetic simulations, for instance, the interpolation of complex-valued fields can lead to spurious solutions. For instance, the average of $1 + 0i$ and $-1 + 0i$ results $0 + 0i$, even if one is expecting to obtain $0 + i$.

This difficulty has been addressed by using an alternative formulation based on the amplitude and phase. For example, in the scenario just described, the average of $1|0$ and $1|\pi$ results in $1|\pi/2$, which is the result we would expect from a physical point of view.

However, the use of this amplitude|phase description still finds difficulties when dealing with the 2π-periodicity. When interpolating a phase close to 2π, e.g. $2\pi - \theta$ ($\theta > 0$, but very small), with another very close too, e.g. $2\pi + \theta'$ ($\theta' > 0$, very small) spurious discontinuities are found. A number being higher than 2π, it reduces to θ', originating the just referred spurious discontinuity: $2\pi - \theta \to \theta'$. Many existing works suggest the use of the so-called *unwrapping* [1–3]. However, typical unwrapping algorithms only work well when the data samples are dense enough, failing in sparse scenarios.

11.1.1 A Simple Phase Unwrapping Procedure

By phase unwrapping we mean the reconstruction of a *physically-meaningful* representation of the phase of a field $E(\theta)$—we highlight the parametric dependence on θ—by adding multiples of 2π to some of its values so as to obtain a continuous function. This step is very important because it determines the number of periods between two successive values of θ, regardless of the interpolation method used. The unwrapped phase is defined as the unique representation of the phase which gives a correct continuous interpolation.

The goal is to find a sequence of integers k_n, $1 \leq n \leq N_\theta$, such that the unwrapped phase ϕ verifies: $\forall n \in [1, N_\theta]$, $\phi_n = \text{Arg } E(\theta_n) + 2k_n\pi$, where Arg is the principal value of the phase in the interval $]-\pi, \pi]$.

Since the problem is ill-posed, a regularization is needed in order to find the correct sequence of integers. The proposed solution is to assume that the derivative of the phase does not vary too much, or, to put it another way, that the second derivative is small. This hypothesis leads to a minimization of the variation of the derivative using a finite differences scheme to compute sequentially the values of the unwrapped phase [4]:

$$k_1 = 0$$
$$k_2 = \underset{k \in \mathbb{Z}}{\text{argmin}} \; |\text{Arg}(E_z(\theta_2)) + 2k\pi - \phi_1|$$
$$\forall n \geq 3, \; k_n = \underset{k \in \mathbb{Z}}{\text{argmin}} \; \left| \frac{\text{Arg}(E_z(\theta_{n+1})) + 2k\pi - \phi_n}{\theta_{n+1} - \theta_n} - \frac{\phi_n - \phi_{n-1}}{\theta_n - \theta_{n-1}} \right|$$

Note that θ_2 must be chosen close enough to θ_1 to ensure $|\phi_2 - \phi_1| < \pi$.

11.2 Interpolating Reduced Bases: The Grassmann Manifold

An important question when defining reduced bases associated with parametric physics concerns the possibility of constructing local bases around a given choice of the model parameters, and then for any other parameters choice interpolating the bases related to the closest neighbors (assuming all of them are defined in the same vector space $\mathbb{R}^n$ (n scaling with the number of nodes in standard nodal-based discretization techniques) and spanning the same (reduced) subspace $\mathbb{R}^p$, with $p \ll n$).

However usual interpolations fail to accomplish the task. To illustrate it, consider a basis $\mathcal{B}_1$ in $\mathbb{R}^2$ composed by vectors $\{(1, 0), (0, 1)\}$ assumed associated to a value of the model parameter p, i.e. $\mathcal{B}(p = 1)$, and a second basis, this time, related to $p = 2$, $\mathcal{B}_2 = \{(-1, 0), (0, -1)\}$ (a simple rotation of the first basis of an angle π).

11.2 Interpolating Reduced Bases: The Grassmann Manifold

Now, even if one could expect for $p = 1.5$ the basis $\{(0, 1), (-1, 0)\}$, the most usual (and naive) linear interpolation between both leads to $\{(0, 0), (0, 0)\}$.

A general route to avoid that issue concerns the formulation of the basis interpolation within an appropriate geometrical framework, involving the concept of manifold $\mathcal{M}$.

As we are interested in manipulating reduced bases composed of p vectors in $\mathbb{R}^n$, we first consider the space of all p-dimensional subspaces of $\mathbb{R}^n$, defined by matrices $Y \in \mathbb{R}^{n \times p}$ with $\text{Rank}(Y) = p$, also called *non-compact Stiefel manifold*, $\mathcal{ST}(p, n)$. Another usual Stiefel manifold is the one composed by all the n × p orthonormal matrices, that represents an embedded sub-manifold of $\mathbb{R}^{n \times p}$.

The *Grassmann manifold* [5–8] that we are introducing now represents a quotient manifold, which requires the partition of matrices into classes of *equivalent* elements representing the same object (an equivalence relation must be reflexive, symmetric and transitive).

Given $Y \in \mathcal{ST}(p, n)$, its span is not unique, it is composed of YM, with M any p × p invertible matrix, defining the set GL_p. Thus, we can define the *Grassmann manifold*, $\mathcal{G}(p, n)$, as the quoitient space $\mathcal{ST}(p, n)/\text{GL}_p$. The equivalence relation involved in the quotient Grassmann manifold is the *span*, that is $X \sim Y$ if and only if $\text{span}(X) = \text{span}(Y)$.

The abstract definition of a manifold $\mathcal{M}$, includes the definition of charts and atlases. The chart $(\mathcal{U}, \varphi)$ is a bijection between a subset $\mathcal{U}$ of $\mathcal{M}$ onto an open subset of $\mathbb{R}^p$, $\varphi(x) \subset \mathbb{R}^p$, that constitutes the coordinates of x. The *atlas* is defined from a collection of charts covering $\mathcal{M}$.

To associate a metric on the manifold $\mathcal{M}$, a first step consists in equipping it with tangent vectors and differential maps. A direct generalization of the directional derivative consists in considering the smooth mapping $\gamma : \mathbb{R} \to \mathcal{M} : t \to \gamma(t)$ called curve, and defining the derivative from

$$\gamma'(t) = \lim_{\tau \to 0} \frac{\gamma(t + \tau) - \gamma(t)}{\tau},$$

however such a choice requires a vector structure to calculate $\gamma(t + \tau) - \gamma(t)$.

To circumvent this issue, we can consider a smooth real-valued function f on $\mathcal{M}$, with $f(\gamma(t))$ a smooth function from $\mathbb{R}$ to $\mathbb{R}$, that allows using the classical derivative to define the tangent vector to the curve γ at $t = 0$, $\dot{\gamma}(0)$. The so-called *tangent bundle* is composed by all the tangent vectors to $\mathcal{M}$. The tangent vectors at a point on the manifold allow defining the so-called tangent space.

The next step, needed for our interpolation purposes, is endowing every tangent space $T_x \mathcal{M}$ with an inner product. The resulting manifold, equipped with a smoothly varying inner product (with the metric described by the metric tensor g) represents the so-called *Riemannian manifold* $(\mathcal{M}, g)$ (a vector space endowed with an inner-product is a particular Riemannian manifold currently called Euclidean space). Associated to the inner product a distance can be defined (positive-definite, symmetric and verifying the triangle inequality).

The *geodesic* $t \to Y(t)$, with $Y(0) = Y_0$ (chosen orthonormal) and $\dot{Y}(0) \in T_{Y_0}\mathcal{G}(p,n)$,

$$Y(t) = Y_0 V \cos(\Sigma t) + U \sin(\Sigma t),$$

with $U\Sigma V$ the thin-SVD decomposition of tangent space representation, $\dot{Y}(0)$. The so-called exponential map results $Y(1)$.

These results will be used later in this work.

References

1. A.P. Shanker, H. Zebker, Edgelist phase unwrapping algorithm for time series InSAR analysis. J. Optical Soc. Am. A **27**(3), 605–612 (2010)
2. M. Costantini, F. Malvarosa, F. Minati, A general formulation for redundant integration of finite differences and phase unwrapping on a sparse multidimensional domain. IEEE Trans. Geosci. Remote Sens. **50**(3), 758–768 (2012)
3. W. Ben Abdallah, R. Abdelfattah, A generalized form of the InSAR phase unwrapping problem based on a compressed sensing technique, in *IEEE International Conference on Image Processing (ICIP)*, Quebec City, QC (2015), pp. 3225–3229
4. S. Vermiglio, V. Champaney, A. Sancarlos, F. Daim, J.C. Kedzia, J.L. Duval, P. Diez, F. Chinesta, Parametric electromagnetic analysis of radar-based advanced driver assistant systems. Sensors **19**, 5686 (2020)
5. E. Begelfor, M. Werman. Affine invariance revisited, in *2006 IEEE Computer Society Conference on Computer Vision and Pattern Recognition (CVPR'06)*, New York, NY, USA (2006), pp. 2087–2094
6. P.A. Absil, R. Mahony, R. Sepulchre, Riemannian geometry of Grassmann manifolds with a view on algorithmic computation. Acta Applicandae Mathematicae **80**(2), 199–220 (2004)
7. P.A. Absil, R. Mahony, R. Sepulchre, *Optimization Algorithms on Matrix Manifolds* (Princeton University Press, 2008)
8. D. Amsallem, C. Farhat, Interpolation method for adapting reduced-order models and application to aeroelasticity. AIAA J. **46**(7), 1803–1813 (2008)

Open Access This chapter is licensed under the terms of the Creative Commons Attribution-NonCommercial-NoDerivatives 4.0 International License (http://creativecommons.org/licenses/by-nc-nd/4.0/), which permits any noncommercial use, sharing, distribution and reproduction in any medium or format, as long as you give appropriate credit to the original author(s) and the source, provide a link to the Creative Commons license and indicate if you modified the licensed material. You do not have permission under this license to share adapted material derived from this chapter or parts of it.

The images or other third party material in this chapter are included in the chapter's Creative Commons license, unless indicated otherwise in a credit line to the material. If material is not included in the chapter's Creative Commons license and your intended use is not permitted by statutory regulation or exceeds the permitted use, you will need to obtain permission directly from the copyright holder.

Chapter 12
Random Variables: Probability, Statistics and Bayesian Learning

Random variables can be discrete or continuous. Their associated probability distributions and statistical inference are major protagonists in data-science and also in Bayesian Learning, BL, a very timely topic due to the consideration of existing knowledge (priors) to determine posterior distributions, performing data-assimilation to enrich the knowledge. Some basic readings on probabilities, statistical inference and random processes, for completing the introduction addressed below are [1, 2].

12.1 Probabilities

A σ-algebra on a set X, is a collection of subsets Σ of X, including itself, closed under the complement, and closed under countable unions (a set is closed when any operation in it results in a member of it). The pair (X, Σ) defines a Borel space. The σ-algebra constitutes the foundation of probability theory.

A measure on X is a function that assigns a positive number to the subsets of X.

12.1.1 Discrete Probability Distributions

If we denote by Ω the set of all possible outcomes, if $x \in \Omega$, then the probability $p(x) \in [0, 1]$, with $\sum_{x \in \Omega} p(x) = 1$, and for a given subset S, $P(S) = \sum_{x \in S} p(x)$.

A discrete distribution becomes fully characterized as soon as $p(x)$ is know. This intrinsic probability determines the difference between the different discrete random variables encountered in physical processes, among them, and to mention a few, the discrete uniform, Bernoulli, binomial, negative binomial, Poisson and the geometric distributions.

12.1.2 Continuous Probability Distributions

We define the probability density function as $p(x)$ and the distribution function as $F(x) = \int_{-\infty}^{x} p(x)dx$, with $\lim_{x \to -\infty} F(x) = 0$ and $\lim_{x \to +\infty} F(x) = 1$.

Among the numerous continuous distributions some examples are the uniform, normal, exponential, gamma and beta distributions.

12.1.3 Moments

A soon as the intrinsic probability distribution is given, all the statistical moments can be obtained straightforward:

- The expectation: $\mathbb{E}(X)$
- The variance: $\text{Var}(X) = \mathbb{E}((X - \mathbb{E}(X))^2)$;
- The covariance: For two random variables X and Y, $\text{Cov}(X, Y) = \mathbb{E}((X - \mathbb{E}(X))(Y - \mathbb{E}(Y)))$.

The *central limit theorem* states that the sum of n independent random variables of any nature, with finite mean μ and variance σ^2, converges towards a normal distribution $\mathcal{N}(n\mu, n\sigma^2)$.

12.1.4 Other General Properties

- $P(\overline{A}) = 1 - P(A)$
- If $A \subset B$, $P(A) \leq P(B)$;
- $P(A \cup B) = P(A) + P(B) - P(A \cap B)$;
- $P(A \cap B) = P(B)P(A|B)$ (if A and B are independent $P(A|B) = P(A)$ or $P(A \cap B) = P(A)P(B)$).
- The Bayes rule:

$$P(B_i|A) = \frac{P(A|B_i)P(B_i)}{P(A)} = \frac{P(A|B_i)P(B_i)}{\sum_j P(B_j)P(A|B_j)}.$$

12.2 Random Vectors

Random vectors are composed of a series of random variables. Their distribution concerns the joint distribution, from which moments can be derived straightforward. It allows defining the marginal variables as illustrated below for a two-dimensional random vector (X, Y), with joint probability distribution $p(x, y)$, that allows calculating the two marginal distributions $p_X(x)$ and $p_Y(y)$ from:

$$p_X(x) = \int_{-\infty}^{\infty} p(x, y) dy,$$

and

$$p_Y(y) = \int_{-\infty}^{\infty} p(x, y) dx,$$

respectively.

12.3 Statistical Inference

12.3.1 Point Estimate

We consider a random variable X, with $\mathbb{E}(X) = \mu$ and $\text{Var}(X) = \sigma^2$, and some events $X_1, X_2, \ldots$:

- The mean value $\overline{X}_n = \frac{1}{n} \sum_{i=1}^{n} X_i$ is an unbiased and convergent estimator of μ;
- If μ is assumed known, $\frac{1}{n} \sum_{i=1}^{n} (X_i - \mu)^2$ represents an unbiased and convergent estimator of σ^2;
- The empirical variance $\frac{1}{n} \sum_{i=1}^{n} (X_i - \overline{X}_n)^2$ is a biased (asymptotically unbiased) estimator of the variance σ^2;
- The empirical variance can be corrected to represent an unbiased estimator: $\frac{1}{n-1} \sum_{i=1}^{n} (X_i - \overline{X}_n)^2$.

12.3.2 Interval Estimate

For the normal distribution $\mathcal{N}(\mu, \sigma^2)$, the confidence interval, of confidence level C, with $\alpha = 1 - C$, is

- When the variance is known, σ^2, then we can compute z such that $z = F^{-1}(1 - \alpha/2)$, from which the confidence interval reads

$$\left(\overline{X}_n - z \frac{\sigma}{\sqrt{n}}, \overline{X}_n + z \frac{\sigma}{\sqrt{n}} \right);$$

- When the variance is unknown, the corrected empirical variance is used instead of σ, and z is obtained from the t-Student distribution.

12.3.3 Statistical Tests

A statistical hypothesis is testable based on observations modeled as random variables. To test normality there is a panoply of available and well experienced tests, among them: (i) D'Agostino's K-squared test; (ii) Jarque-Bera test; (iii) Anderson-Darling test, (iv) Cramer-von Mises criterion, (v) Kolmogorov-Smirnov test, (vi) Lilliefors test, (vii) Shapiro-Wilk test, (viii) Pearson's chi-squared test, ...

Tests based on Bayes compare priors and posterior distributions, using the very useful Kullback-Leibler divergences to compare probability distributions.

12.4 Bayesian Learning

Bayesian inference or Bayesian learning [3] is a very powerful tool for data assimilation or to incorporate existing knowledge (prior) in machine learning techniques.

Let us consider:

- A quantity of interest u that depends on a parameter value θ leading to $p(u|\theta)$. Model Order Reduction, addressed later, facilitates its construction.
- A prior knowledge on the parameter distribution $p(\theta)$
- A datum u

We would like to estimate from the observation u the most probable value of θ, i.e., $p(\theta|u)$. By virtue of Bayes, we can write

$$p(\theta|u) = \frac{p(u|\theta)p(\theta)}{p(u)}.$$

Since, in general, one looks for the θ that maximizes $p(\theta|u)$, the numerator suffices to extract the searched value, which is expressed as: posterior is proportional to the prior times the likelihood. When the prior is uniform, posterior coincides, as expected, with the likelihood.

12.4.1 Naive Bayes

We assume a vector with n features $\mathbf{x} = (x_1, \ldots x_n)$ from which the class c_k, among the K available is being inferred from $\mathbf{x}$.

Bayes writes

$$p(c_k|\mathbf{x}) = \frac{p(c_k)p(\mathbf{x}|c_k)}{p(\mathbf{x})},$$

however, the numerator can be rewritten as $p(x_1, \ldots, x_n, c_k)$, therefore assuming the features mutually independent, conditional on c_k, results in

$$p(x_1, \ldots, x_n, c_k) = p(c_k) \prod_{i=1}^{n} p(x_i | c_k).$$

References

1. F.M. Dekking, C. Kraaikamp, H.P. Lopuhaa, L.E. Meester, *A Modern Introduction to Probability and Statistics. Understanding Why and How* (Springer, 2005)
2. H. Pishro-Nik, *Introduction to Probability, Statistics, and Random Processes* (Kappa Research LLC, 2014)
3. J.O. Berger, *Statistical Decision Theory and Bayesian Analysis*. Springer Series in Statistics (1985)

Open Access This chapter is licensed under the terms of the Creative Commons Attribution-NonCommercial-NoDerivatives 4.0 International License (http://creativecommons.org/licenses/by-nc-nd/4.0/), which permits any noncommercial use, sharing, distribution and reproduction in any medium or format, as long as you give appropriate credit to the original author(s) and the source, provide a link to the Creative Commons license and indicate if you modified the licensed material. You do not have permission under this license to share adapted material derived from this chapter or parts of it.

The images or other third party material in this chapter are included in the chapter's Creative Commons license, unless indicated otherwise in a credit line to the material. If material is not included in the chapter's Creative Commons license and your intended use is not permitted by statutory regulation or exceeds the permitted use, you will need to obtain permission directly from the copyright holder.

Chapter 13
Random Variables: Polynomial Chaos, PC

This section revisits the foundations of polynomial chaos, of major interest for expressing random fields and performing uncertainty quantification and propagation [1, 2].

13.1 Functions of Random Variables

13.1.1 The Univariate Case

We will denote by X the random variable of interest and Ξ the random variable in terms of which X will be expressed, also called *germ*. The choice of the germ is a modeling choice, e.g. uniform, gaussian, ...

To represent X a usual route consists in expressing it from $X = f(\Xi)$, and finding the adequate $f(\bullet)$, such that for the chosen germ Ξ we obtain the required distribution for X.

13.1.2 Polynomial Chaos, PC

The first route consists in expanding the function f in a polynomial series. Orthogonal polynomials (with respect to the chosen germ) represent an appealing choice, as proved later. Thus, we define the scalar product

$$\langle h_1(\xi), h_2(\xi) \rangle = \int h_1(\xi) h_2(\xi) p_\Xi(\xi) d\xi,$$

where $p_\Xi(\xi)$ is the probability density function of Ξ.

The considered basis is composed of $\Psi_0 = 1, \Psi_1(\Xi), ...$, with $\Psi_n(\Xi)$ a polynomial of order n, verifying $\langle \Psi_i, \Psi_j \rangle = 0$ if $i \neq j$.

Since $\Psi_0 = 1$, all the higher order random variables $\Psi_i(\Xi), i \geq 1$ have zero mean (Ψ_i being orthogonal to Ψ_0). Moreover $\langle \Psi_i, \Psi_i \rangle$ represents the variance of Ψ_i, the covariances being zero (uncorrelated random variables) by virtue of orthogonality.

By construction, (i) a germ uniformly distributed leads to Legendre polynomial; (ii) standard normally distributed to the Hermite polynomials, (iii) an exponential random variable to the Laguerre polynomials, ...

Now, X can be expressed by

$$X = f(\Xi) = \sum_{i=0}^{\infty} x_i \Psi_i(\Xi),$$

with coefficients x_i computable in virtue of the orthogonality, from

$$x_i = \frac{\langle f, \Psi_i \rangle}{\langle \Psi_i, \Psi_i \rangle},$$

expansion called *Polynomial Chaos*.

The truncated expansion counterpart reads

$$X_p = f_p(\Xi) = \sum_{i=0}^{p} x_i \Psi_i(\Xi).$$

13.1.3 The Multivariate Case

Here both X and Ξ are vectors: $\mathbf{X}$ and $\mathbf{\Xi}$. By using the index $\mathbf{i} = (i_1, \ldots, i_m)$,

$$\Psi_{\mathbf{i}}(\mathbf{\Xi}) = \prod_{j=1}^{m} \Psi_{i_j}(\Xi_j).$$

13.2 Polynomial Chaos Expansion for Random Fields

Now, we consider (without loss of generality assumed univariate) the random field $X(t)$ that can be approximated as

$$X(t) \approx \sum_{i=0}^{p} x_i(t) \Psi_i(\Xi),$$

13.3 Uncertainty Propagation, UP

that generalizes the Karhunen-Loeve expansion

$$X(t) = \mathbb{E}\{X(t)\} + \sum_{i=1}^{\infty} \sqrt{\lambda_i} \phi_i(t) \Xi_i,$$

with Ξ_i uncorrelated with zero mean.

13.3 Uncertainty Propagation, UP

Here we consider a model $Y = \eta(X)$, again without loss of generality, univariate, with the input and output, X and Y random variables. Uncertainty propagation aims at propagating the known input uncertainty to the output.

With the input characterized

$$X = \sum_{j=0}^{p} x_j \Psi_j(\Xi),$$

we assume the output expressed in a similar way,

$$Y = \sum_{j=0}^{p} y_j \Psi_j(\Xi),$$

with y_j unknown.

Thus, uncertainty propagation consists in solving

$$\sum_{j=0}^{p} y_j \Psi_j(\Xi) = \eta \left(\sum_{j=0}^{p} x_j \Psi_j(\Xi) \right),$$

to compute coefficients y_j.

13.3.1 Intrusive Procedure

The projection

$$\left\langle \sum_{j=0}^{p} y_j \Psi_j(\Xi), \Psi_k \right\rangle = \left\langle \eta \left(\sum_{j=0}^{p} x_j \Psi_j(\Xi) \right), \Psi_k \right\rangle, \quad k = 0, \ldots, p,$$

results in a series of $p+1$ equations to calculate the $p+1$ unknown coefficients y_i, $i = 0, 1, \ldots, p$. However, the complexity of function $\eta(\bullet)$ makes difficult the evaluation of the right hand member.

13.3.2 Non-intrusive Procedure Based on PC Expansion

The non-intrusive route considers a set of M values of Ξ, Ξ_m, from which result the corresponding X_m that allow evaluating the associated model output $Y_m = \eta(X_m)$, $\forall m$. Thus, the output expansion coefficients read

$$y_i = \frac{\langle Y, \Psi_i \rangle}{\langle \Psi_i, \Psi_i \rangle},$$

where the integral in the numerator is computed by using an appropriate quadrature, i.e.

$$\langle Y, \Psi_i \rangle = \sum_{m=1}^{M} \omega_m Y_m \Psi_i(\Xi_m) p_\Xi(\Xi_m).$$

Of course, as soon as the output is available in a large number of points, and even if the reconstruction of the distribution itself would require much more data, the calculation of the distribution moments requires much less data, and then can be also computed from the M available data:

$$\mathbb{E}(Y) = y_0,$$

as well as the higher order moments:

$$\mathbb{E}(Y^2) = \sum_{j=0}^{p} y_j^2 \langle \Psi_j, \Psi_j \rangle.$$

13.3.3 Non-intrusive Procedure Based on Regression

The main difficulty related to the just introduced procedure concerns the computational cost to evaluate the output Y_m, related to the input $X_m \equiv X(\Xi_m)$, when M increases. The offline construction of a surrogate model (a parametric solution in the PGD framework addressed later) $Y = Y(X)$ facilitates the computational efficiency.

Thus, the results Y_m are used to construct a regression, for example:

$$Y = \sum_{j=0}^{p} y_j \Psi_j(\Xi),$$

with the coefficients y_j estimated by using least-squares or some other inferential method.

References

1. R. Ghanem, P. Spanos, *Stochastic Finite Elements: A Spectral Approach* (Springer, 1991)
2. A. O'Hagan, *Polynomial Chaos: A Tutorial and Critique from a Statistician Perspective* (2013)

Open Access This chapter is licensed under the terms of the Creative Commons Attribution-NonCommercial-NoDerivatives 4.0 International License (http://creativecommons.org/licenses/by-nc-nd/4.0/), which permits any noncommercial use, sharing, distribution and reproduction in any medium or format, as long as you give appropriate credit to the original author(s) and the source, provide a link to the Creative Commons license and indicate if you modified the licensed material. You do not have permission under this license to share adapted material derived from this chapter or parts of it.

The images or other third party material in this chapter are included in the chapter's Creative Commons license, unless indicated otherwise in a credit line to the material. If material is not included in the chapter's Creative Commons license and your intended use is not permitted by statutory regulation or exceeds the permitted use, you will need to obtain permission directly from the copyright holder.

Chapter 14
Radom Variables: Gaussian Processes, GP

GPs provide an alternative approach to the regression just addressed [1, 2]. If we consider the linear regression $Y = \theta_0 + \theta_1 X$, Bayesian linear regression provides a probabilistic way to estimate these coefficients from the collected data, within a parametric framework. On the contrary GP represents a non-parametric approach, finding over all possible functions $f(X)$ those consistent with the available data, that is, something like we have an infinity of parameters. Like all the Bayesian methods, it starts with some priors that are updated from the data to avoid considering *all possible functions*.

14.1 Multivariate Gaussian Distribution

We assume the data vector $\mathbf{X} = [X_1, \ldots, X_M]$ with each component being a random variable normally distributed. Thus, $\mathbf{X}$ can be described as $\mathcal{N}(\boldsymbol{\mu}, \boldsymbol{\Sigma})$, $\boldsymbol{\mu}$ being the $M \times 1$ vector composed by the mean of each random variable X_i, and $\boldsymbol{\Sigma}$ the $M \times M$ covariance matrix $\Sigma_{ij} = \mathbb{E}\{(X_i - \mu_i)(X_j - \mu_j)\}$.

Gaussian distributions are very common in physics due to the fact that the average of any random process follows a normal distribution according to the *central limit theorem*. This distribution has remarkable properties that we are summarizing here. If we assume a bivariate Gaussian distribution with joint probability $p(X, Y)$

$$p(X, Y) \sim \mathcal{N}(\boldsymbol{\mu}, \boldsymbol{\Sigma}),$$

$\boldsymbol{\mu}^T = (\mu_X, \mu_Y)$ and the symmetric covariance matrix expressed by

$$\boldsymbol{\Sigma} = \begin{pmatrix} \Sigma_{XX} & \Sigma_{XY} \\ \Sigma_{YX} & \Sigma_{YY} \end{pmatrix},$$

with $\Sigma_{XY} = \Sigma_{YX}$, the marginal probability distributions are also normally distributed: $p_X(X) = \int_Y p(x, y) dy \sim \mathcal{N}(\mu_x, \Sigma_{XX})$, and $p_Y(Y) = \int_X p(x, y) dx \sim \mathcal{N}(\mu_y, \Sigma_{YY})$.

The conditional probability $p(X|Y)$ (analogously, $p(Y|X)$) can be computed by using the relation: $p(X, Y) = p(X|Y) p_Y(Y)$ (respectively, $p(X, Y) = p(Y|X) p_X(X)$), from which we obtain:

$$p(X|Y) \sim \mathcal{N}(\mu_x + \Sigma_{XY} \Sigma_{YY}^{-1} (Y - \mu_Y), \Sigma_{XX} - \Sigma_{XY} \Sigma_{YY}^{-1} \Sigma_{YX}),$$

and

$$p(Y|X) \sim \mathcal{N}(\mu_y + \Sigma_{YX} \Sigma_{XX}^{-1} (X - \mu_X), \Sigma_{YY} - \Sigma_{YX} \Sigma_{XX}^{-1} \Sigma_{XY}).$$

It can be seen that only the means change with the available information, respectively Y and X.

In the previous developments we considered perfect measurements, but such a scenario is unrealistic. Usually, an error is introduced, modeled from the normal distribution $\mathcal{N}(0, \Upsilon^2)$, at each training point, implying the consideration of $\Sigma_{YY} + \Upsilon^2 \mathbf{I}$.

14.2 Gaussian Process

We assume a M-dimensional multivariate distribution, with the covariance given by a positive-definite kernel $\Sigma_{ij} = \kappa(X_i, X_j)$. Kernels can be stationary or non-stationary. The former only depends on the relative distance between the i and j data-points, whereas the last can also depend on the absolute location.

We assume that m ($m < M$) data is available, constituting Y (training data) and we would like to estimate the value at the $M - m$ remaining points X. Training data allows updating the posterior distribution according to the conditional distribution $P(X|Y)$, that allows computing averages and confidence intervals.

The approximation technique called *Kriging* (addressed in Chap. 19) is based on the just introduced concepts and procedures.

References

1. C.E. Rasmussen, C.K.I. Williams, *Gaussian Processes for Machine Learning* (MIT Press, 2006)
2. J. Gortler, R. Kehlbeck, O. Deussen, *A Visual Exploration of Gaussian Processes* (2019), https://doi.org/10.23915/distill.00017

Open Access This chapter is licensed under the terms of the Creative Commons Attribution-NonCommercial-NoDerivatives 4.0 International License (http://creativecommons.org/licenses/by-nc-nd/4.0/), which permits any noncommercial use, sharing, distribution and reproduction in any medium or format, as long as you give appropriate credit to the original author(s) and the source, provide a link to the Creative Commons license and indicate if you modified the licensed material. You do not have permission under this license to share adapted material derived from this chapter or parts of it.

The images or other third party material in this chapter are included in the chapter's Creative Commons license, unless indicated otherwise in a credit line to the material. If material is not included in the chapter's Creative Commons license and your intended use is not permitted by statutory regulation or exceeds the permitted use, you will need to obtain permission directly from the copyright holder.

Chapter 15
Analysis of Variance, ANOVA

15.1 ANOVA Decomposition

A function $f(\mathbf{x})$, assumed depending on D random variables, $x_1, \ldots, x_D$, can be decomposed as [1]

$$f(\mathbf{x}) = f_0 + \sum_{i=1}^{D} f_i(x_i) + \sum_{i_1=1}^{D} \sum_{i_2=i_1}^{D} f_{i_1,i_2}(x_{i_1}, x_{i_2}) + \cdots + f_{1,2,\ldots,D}(x_1, x_2, \ldots, x_D),$$

that, with the property of

$$\mathbb{E}_i(f_{i_1,\ldots,i_d}(x_{i_1}, \ldots, x_{i_d})) = 0,$$

with $\mathbb{E}_i$ the expectation with respect to any coordinate i in the set $(i_1, \ldots, i_d)$, $1 \le d \le D$, results in the orthogonality of functions involved in the previous decomposition.

To prove it, consider for example a simple 2D case with $f_{x,y}(x, y)$ and $f_x(x)$. Thus, with $\mathbb{E}_x(f_x(x)) = 0$, $\mathbb{E}_x(f_{x,y}(x, y)) = 0$ and $\mathbb{E}_y(f_{x,y}(x, y)) = 0$, we have $\mathbb{E}_{x,y}(f_{x,y}(x, y) f_x(x)) = \mathbb{E}_x \{\mathbb{E}_y(f_{x,y}(x, y)) f_x(x)\} = 0$.

The number of functions involved in the decomposition (without considering the constant term) is $2^D - 1$, and they can be parametrized by the integer s, $s = 1, \ldots, 2^D - 1$.

The different functions involved in the ANOVA decomposition can be expressed from expectations according to:

$$\begin{cases} \mathbb{E}(f(\mathbf{x})) = f_0 \\ \mathbb{E}(f(\mathbf{x}|x_i)) = f_i(x_i) + f_0 \\ \mathbb{E}(f(\mathbf{x}|x_i, x_j)) = f_{i,j}(x_i, x_j) + f_i(x_i) + f_j(x_j) + f_0 \\ \vdots \end{cases}$$

where $\mathbb{E}(f(\mathbf{x}|x_i))$ refers to the integration on all the variables except x_i.

15.2 Sensitivity Analysis: Sobol Coefficients

The variance of $f(\mathbf{x})$, $\text{Var}(f(\mathbf{x}))$, taking into account the orthogonality of the functions involved in the ANOVA decomposition, reads

$$\text{Var}(f(\mathbf{x})) = \sum_{s=1}^{2^D-1} (f_s(\mathbf{x}_s))^2 = \sum_{s=1}^{2^D-1} \text{Var}_s,$$

from which the so-called Sobol sensitivity coefficients $\mathcal{S}_s$ read

$$\mathcal{S}_s = \frac{\text{Var}_s}{\text{Var}(f(\mathbf{x}))}, \quad s = 1, ..., D.$$

15.3 Anchored ANOVA

When addressing multidimensional settings the calculation of the multidimensional expectations becomes computationally expensive. For alleviating this difficulty we introduce the so-called anchor point $\mathbf{c}$ such that $f_0 = f(\mathbf{c})$ [2]. Then, the expectations involved by the ANOVA decomposition are replaced by $f(\mathbf{c}|\mathbf{x}_s)$, that is, the particularization of the function at the anchor point, except for those coordinates involved in $\mathbf{x}_s$.

References

1. G.E.B. Archer, A. Saltelli, I.M. Sobol, Sensitivity measures, anova-like techniques and the use of bootstrap. J. Stat. Comput. Simul. **58**, 99–120 (1997)
2. K. Tang, P.M. Congedo, R. Abgrall, Sensitivity analysis using anchored ANOVA expansion and high order moments computation. Research Report RR-8531, INRIA (2014)

References

Open Access This chapter is licensed under the terms of the Creative Commons Attribution-NonCommercial-NoDerivatives 4.0 International License (http://creativecommons.org/licenses/by-nc-nd/4.0/), which permits any noncommercial use, sharing, distribution and reproduction in any medium or format, as long as you give appropriate credit to the original author(s) and the source, provide a link to the Creative Commons license and indicate if you modified the licensed material. You do not have permission under this license to share adapted material derived from this chapter or parts of it.

The images or other third party material in this chapter are included in the chapter's Creative Commons license, unless indicated otherwise in a credit line to the material. If material is not included in the chapter's Creative Commons license and your intended use is not permitted by statutory regulation or exceeds the permitted use, you will need to obtain permission directly from the copyright holder.

Part II
Around Learning

Chapter 16
Data Classification and Clustering

The present section revisits some usual clustering (unsupervised) and classification (supervised) techniques. More advanced procedures will be addressed in the next sections.

16.1 Unsupervised Clustering: k-Means

k-means is one of the most popular clustering techniques [1, 2]. k-means groups a set of data into a number or groups, the so-called clusters, such that, members of a cluster are closer to all the members of that cluster than to any other data belonging to another cluster.

If we consider a set of high-dimensional data, consisting of vectors $(\mathbf{x}_1, \mathbf{x}_2, \ldots, \mathbf{x}_M)$, $\mathbf{x}_i \in \mathbb{R}^D$, $i = 1, \ldots, M$. Our objective is distributing these M data into k sets ($k \leq M$), $\mathcal{S} = \{S_1, S_2, \ldots, S_k\}$, such that

$$\mathcal{S} = \arg\min_{\mathcal{S}^*} \sum_{i=1}^{k} \sum_{\mathbf{x} \in S_i^*} \|\mathbf{x} - \boldsymbol{\mu}_i\|^2,$$

where $\boldsymbol{\mu}_i$ is the mean of each cluster.

16.2 Supervised Data Classification

16.2.1 Support Vector Machines, SVM

Here, considering labeled data in $\mathbb{R}^D$, the objective consists of finding the frontiers between the different data classes.

Fig. 16.1 Sketch of a support vector machine. The algorithm finds the hyperplane whose *margin* is maximum

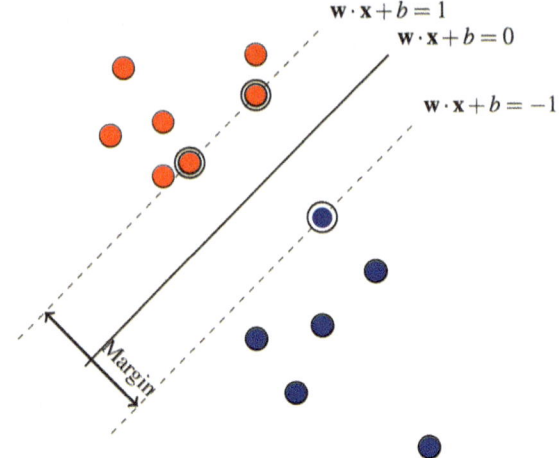

Support vector machines look for hyperplanes (in the case of linear classification) in the data space $\mathbb{R}^D$ maximizing the margin, that is, the distance between the data and those hyperplanes to maximize performances and robustness in the classification. The nonlinear case proceeds in a similar way, and employs kernels [3].

With the data labels y_i, taking the values $+1$ or -1, that is, considering two classes, the red and blue points in Fig. 16.1, classification of data $(\mathbf{x}_1, y_1), \ldots, (\mathbf{x}_M, y_M)$, looks for the hyperplane (one-dimensional in the case presented in Fig. 16.1) separating the data into two sets, while maximizing the distance between the data and the searched hyperplane.

In Fig. 16.1 the frontier reads $\mathbf{w} \cdot \mathbf{x} - b = 0$, where $\mathbf{w}$ is the normal vector to the hyperplane. The distance between the hyperplanes defined by $\mathbf{w} \cdot \mathbf{x} - b = \pm 1$, represented as dashed lines in Fig. 16.1, is $2/\|\mathbf{w}\|$. Thus, the margin maximization implies minimizing $\|\mathbf{w}\|$, with the separation constraint

$$y_i(\mathbf{w} \cdot \mathbf{x}_i - b) \geq 1, \quad \forall i,\ 1 \leq i \leq M.$$

The constrained minimization allows obtaining both, $\mathbf{w}$ and b, from which the classifier reads:

$$\mathbf{x} \mapsto \text{sign}\,(\mathbf{w} \cdot \mathbf{x} - b).$$

16.2.2 Decision Trees and Random Forest

From the data set $\mathcal{D} = \{((\mathbf{x}_i, y_i), i = 1, 2, \ldots, M\}$, the objective consists of predicting y associated with a datum $\mathbf{x}$. The output y can refer the class (classification) or a real value in the case of regression [4, 5].

A Decision Tree consists of nodes and branches. Data is divided into two branches at each node, depending on the response to an appropriate question: true or false. Thus, the problem reduces to find the right questions formulated at the right place. For that purpose, a grade of the so-called *impurity* is attached to each node.

The node impurity is usually quantified by using the Gini impurity criterion. Consider a set of C different classes, $c = 1, 2, \ldots, C$, with p_c the fraction of elements at the node belonging to the class c. The Gini impurity reads

$$I(p_1, \ldots, p_C) = \sum_{c=1}^{C} p_c(1 - p_c),$$

that quantifies the probability to be assigned to the correct class times the one of being assigned to the wrong class.

The Gini impurity is calculated at each node, before and after separation, whose difference leads to the so-called *information gain*. By comparing the information gain associated with each possible question, the one providing the maximum information gain becomes the optimal one.

If we consider different smaller data-sets consisting of $M' < M$ data-points, randomly chosen, a set of trees result, the so-called forest. If a new data **x** is evaluated by using each tree, and the classification predicted by each tree computed, the random forest classification consists of the class that obtains the biggest number of votes for the ensemble of trees.

References

1. J. MacQueen, Some methods for classification and analysis of multivariate observations, in *Proceedings of 5th Berkeley Symposium on Mathematical Statistics and Probability* (University of California Press, 1967), pp. 281–297
2. D. MacKay, Chapter 20. *An Example Inference Task: Clustering. Information Theory, Inference and Learning Algorithms* (Cambridge University Press, 2003), pp. 284–292
3. N. Cristianini, J. Shawe-Taylor, *An Introduction to Support Vector Machines and Other Kernel-based Learning Methods* (Cambridge University Press, New York, 2000)
4. C.W. Kirkwood, Decision Tree Primer (2002). http://creativecommons.org/licenses/by-nc/3.0/
5. L. Breiman, Random Forests. Mach. Learn. **45**, 5–32 (2001)

Open Access This chapter is licensed under the terms of the Creative Commons Attribution-NonCommercial-NoDerivatives 4.0 International License (http://creativecommons.org/licenses/by-nc-nd/4.0/), which permits any noncommercial use, sharing, distribution and reproduction in any medium or format, as long as you give appropriate credit to the original author(s) and the source, provide a link to the Creative Commons license and indicate if you modified the licensed material. You do not have permission under this license to share adapted material derived from this chapter or parts of it.

The images or other third party material in this chapter are included in the chapter's Creative Commons license, unless indicated otherwise in a credit line to the material. If material is not included in the chapter's Creative Commons license and your intended use is not permitted by statutory regulation or exceeds the permitted use, you will need to obtain permission directly from the copyright holder.

Chapter 17
Boosting Algorithms

Boosting is more a generic methodology than as a particular method. It is based on the fact that more than training a single model, one could make use of several *weak* models (machine learning algorithms performing simply better than random procedures), each one trying to compensate the weaknesses of the predecessors.

In what follow we revisits two algorithms: the AdaBoost and the Gradient Boosting.

17.1 AdaBoost

AdaBoost is widely employed in classification, where the application of a sequence of simple classifiers, usually over-performs the use of complex classifiers [1].

If we consider a classifier $\mathcal{C}$, that applied on the data x_i infer the class $\mathcal{C}(x_i) = \pm 1$, and we make use of the function that takes a unit value when the prediction is correct, and zero otherwise, i.e.

$$\mathcal{I}(\mathcal{C}(x)) = \begin{cases} 1 \text{ if } \mathcal{C}(x) = y, \\ 0 \text{ if } \mathcal{C}(x) \neq y \end{cases},$$

where y refers to the class to which datum x belongs, then, the classifier error is defined for the N available data

$$\mathcal{E}(\mathcal{C}, \boldsymbol{\omega}) = \frac{\sum_{i=1}^{N} \omega_i (1 - \mathcal{I}(\mathcal{C}(x_i)))}{\sum_{i=1}^{N} \omega_i}, \qquad (17.4)$$

© The Author(s) 2025
F. Chinesta et al., *A Gentle Introduction to Data, Learning, and Model Order Reduction*, Studies in Big Data 174, https://doi.org/10.1007/978-3-031-87572-4_17

where w_i are the weights (grouped into vector ω) associated to each data, whose calculation is described below.

From M *weak* available classifiers $\mathcal{C}^j$, $j = 1, ..., M$, AdaBoost proceeds by applying sequentially each one to update the weights w_i, $i = 1, ..., N$, while using those weights in the classifiers training, in such a way that higher errors leads to increase the weights to improve the solution where the previous classifier underperforms.

From the simplest initialization consisting of $w_i = 1/N$, the algorithm proceeds as follows:

- Each classifier $\mathcal{C}^j$, $j = 1, ..., M$ performs
    - Use the weak classifier $\mathcal{C}^j$, with the data, labels and weights $x_1, ..., x_N, y_1, ..., y_N$ and $w_1, ..., w_N$;
    - Compute its error $\mathcal{E}(\mathcal{C}^j, \omega)$, using Eq. (17.4) on the trained classifier $\mathcal{C}^j$ predictions;
    - Compute a global coefficient α^j associated to the classifier $\mathcal{C}^j$, such that it increases with the error, for example

    $$\alpha^j = \log\left(\frac{1 - \mathcal{E}(\mathcal{C}^j, \omega)}{\mathcal{E}(\mathcal{C}^j, \omega)}\right);$$

    - Weights updating, looking for higher values where the classifier underperforms, for example by using

    $$w_i \leftarrow w_i \exp\{\alpha^j (1 - \mathcal{I}(x_i))\},$$

    and then performing the normalization ω.

- After finishing the action of the M concatenated classifiers, the label ($k = \pm 1$) prediction results from a weighted voting, for example

$$\mathcal{C}^{\texttt{AdaBoost}}(x) = \sum_{j=1}^{M} \alpha^j \mathcal{C}^j(x).$$

17.2 Gradient Boosting and Its Stochastic Counterpart

In gradient boosting instead of adapting the data weights, is the difference between prediction and ground truth that is used, and boosting is reinterpreted as an optimization problem [2].

At each iteration j, that considers a *weak* learner $\mathcal{C}^j$, a lost is defined $\text{L}(\mathcal{C}^j)$, e.g.

$$\text{L}(\mathcal{C}^j) = \frac{1}{2N} \sum_{i=1}^{N} \left(y_i - \mathcal{C}^j(x_i)\right)^2,$$

whose *neg-gradient* at each data point reads

$$-\frac{\partial L(\mathcal{C}^j)}{\partial \mathcal{C}^j(x_i)} = -\frac{1}{N}(y_i - \mathcal{C}^j(x_i)) \equiv \mathcal{R}_i^j. \qquad (17.5)$$

The algorithm performs at a certain iteration j, from $\mathcal{C}^0$ that can be initialized from the outputs average $\bar{y} = \sum_{i=1}^{N} y_i$, i.e., $\mathcal{C}^0(x_i) = \bar{y}, i = 1, ..., N$, as follows:

- Compute quantities $\mathcal{R}_i^j$ according to Eq. (17.5);
- Use a *weak* learner for $\mathcal{H}^j(x)$ from the data $\{(x_i, \mathcal{R}_i^j), ..., (x_N, \mathcal{R}_N^j)\}$;
- Solve the one-dimensional minimization problem

$$\gamma^j = \arg\min_\gamma \left(\sum_{i=1}^{N} L(\mathcal{C}^j(x_i) + \gamma \mathcal{H}^j(x_i)) \right);$$

- The model updating proceeds according to:

$$\mathcal{C}^{j+1}(x) = \mathcal{C}^j(x) + \gamma^j \mathcal{H}^j(x).$$

A slight but very effective variation of the procedure just described was proposed in [3] where instead of considering at each iteration the full dataset, a random part of it, containing a given fraction of the data is considered in the procedure just described, giving rise to the so-called stochastic gradient boost.

References

1. Y. Freund, R. Schapire, A decision-theoretic generalization of on-line learning and an application to boosting. J. Comput. Syst. Sci. **55**(1), 119–139 (1997)
2. J. Friedman, Greedy function approximation: a gradient boosting machine. Ann. Stat. **29**, 1189–1232 (2000)
3. J. Friedman, Stochastic gradient boosting. Comput. Stat. Data Anal. **38**(4), 367–378 (2002)

Open Access This chapter is licensed under the terms of the Creative Commons Attribution-NonCommercial-NoDerivatives 4.0 International License (http://creativecommons.org/licenses/by-nc-nd/4.0/), which permits any noncommercial use, sharing, distribution and reproduction in any medium or format, as long as you give appropriate credit to the original author(s) and the source, provide a link to the Creative Commons license and indicate if you modified the licensed material. You do not have permission under this license to share adapted material derived from this chapter or parts of it.

The images or other third party material in this chapter are included in the chapter's Creative Commons license, unless indicated otherwise in a credit line to the material. If material is not included in the chapter's Creative Commons license and your intended use is not permitted by statutory regulation or exceeds the permitted use, you will need to obtain permission directly from the copyright holder.

Chapter 18
Learning Modalities

Learners locating somewhere in between supervised and unsupervised learning, are expected being a real revolution in artificial intelligence. For a recent talk of Y. LeCun the interested reader can refer to [1]. Transfer and Reinforced Learning represents at their turn very powerful learning procedures, the former by profiting from some existing knowledge, and the later by reinforcing the learning by using the so-called reward.

18.1 Self-supervised Learning

Self-supervised learning uses the available data for supervising the learning process, circumventing, the tedious labelling task.

To illustrate the procedure, we consider the self-supervised learning applied to infer images rotations. For that, following [2], we consider a set of N training images. A convolutional neural network, CNN, extracts the features of images in an unsupervised manner.

We note by X an image and $F(X)$ the features that results from the CNN applied on X. Now, a transformation (rotation) is applied to the image $g(X|r)$, where r parametrize the applied rotation, that results in X^r, with the features extracted by the CNN expressed by $F(X^r)$.

In the training state, K rotations $r_k, k = 1, ..., K$, are applied to the different images $X_i, i = 1, ..., N$, leading to $X_i^{r_k}$. The parameters θ of the Neural Network, NN, taking as input the features extracted by the CNN, and inferring as output the applied rotation, uses as loss function:

$$\theta = \min_{\theta^*} \frac{1}{N} \sum_{i=1}^{N} \texttt{loss}(X_i, \theta^*),$$

where the loss term is expressed by

$$\text{loss}(X_i, \theta^*) = -\frac{1}{K} \sum_{k=1}^{K} \log \mathcal{R}_{\theta^*}^{r_k}(F(X_i^{r_k})),$$

where $\mathcal{R}_{\theta^*}^{r_k}$ gives the score for a rotation r_k.

18.2 Semi-supervised Learning

In the case of semi-supervised we assume a set of data $\{x_1, x_2, ..., x_D, x_{D+1}, ...x_{D+d}\}$, where we assume that the associated labels are known for the first D data, $\{y_1, ..., y_D\}$, and are unknown for the d remaining data.

Transductive learning looks for inferring the d remaining labels, $\{y_{D+1}, ..., y_{D+d}\}$, whereas *inductive learning* looks for the estimator $y = y(x)$.

Different assumptions are used in this process. Among them, (i) the continuity, that expresses that points that are close (using an adequate metrics) are expected having the same label; (ii) the fact that data tends to group in different clusters; and (iii) that data locate on a manifold of much lower dimension that the embedding space.

In what follows we describe three techniques among all the existing. For more details the interested reader can refer to [3].

- *Generative methods.*

  Generative methods uses Bayes rule. It estimates the joint probability $P(x, y|\theta)$ from $p(y|\theta)p(x|y, \theta)$, where θ represents the parameters considered in the joint probability. The prior distribution can be considered a multinomial over $\mathcal{Y}$ and $p(x, y|\theta)$ could be a multivariate Gaussian.

  Thus, for the labelled data the right choice of the parameters θ will be

  $$\theta = \arg\max_{\theta^*} \left\{ \log \left(p(\{x_i, y_i\}_{i=1}^{D} | \theta^*) \right) \right\},$$

  however, the knowledge of the unlabeled data can also be used for improving the estimation. For that purpose, we consider

  $$\log \left(p(\{x_i\}_{i=D+1}^{D+d} | \theta) \right) = \sum_{i=D+1}^{D+d} \log \left(\sum_{y \in \mathcal{Y}} p(x_i, y|\theta) \right),$$

  and then both combined according to

  $$\theta = \arg\max_{\theta^*} \left\{ \log \left(p(\{x_i, y_i\}_{i=1}^{D} | \theta^*) \right) + \lambda \log \left(p(\{x_i\}_{i=D+1}^{D+d} | \theta^*) \right) \right\}.$$

- *Semi-supervised SVM.*

    SVM finds $f(x)$ such that $f(x) = 0$ separates the class of data. Now, to the usual labelled (supervised) formulation

    $$f = \underset{f^*}{\arg\min} \left\{ \frac{1}{D} \sum_{i=1}^{D} \max(1 - y_i f^*(x_i), 0) + \lambda_1 \|f^*\|^2 \right\}, \tag{18.6}$$

    to which we should add the fact that $f(x) = 0$ should locate in the region of low density of unlabeled data, from the use of a second penalty

    $$\lambda_2 \frac{1}{d} \sum_{D+1}^{D+d} \max(1 - |f^*(x)|, 0),$$

    to be added to Eq. (18.6).
- *Graphs based methods.*

    A graph consists of a set of vertices V and a set of edges E joining the vertices. Thus the graph is given by $\{V, E\}$. Each edge connecting nodes x_i and x_j is affected by a weight that reflects the distance between the two data, as a Gaussian weight

    $$w_{ij} = \exp(\|x_i - x_j\|/\sigma^2),$$

    even if many other possibilities exist.
    The energy of the graph is given by

    $$\mathcal{U}(f) = \sum_{i,j=1}^{D+d} w_{ij}(f(x_i) - f(x_j)).$$

    Thus, the function f should fit the labelled data and minimize the energy of the graph:

    $$f = \underset{f^*}{\arg\min} \left\{ \frac{1}{D} \sum_{i=1}^{D} c(f^*(x_i), y_i) + \lambda_1 \|f^*\|^2 + \lambda_2 \mathcal{U}(f^*) \right\},$$

    where $c(f^*(x_i), y_i)$ is a convex function enforcing the data fit.

18.3 Transfer Learning

Transfer learning aims at transferring knowledge from one domain to another, inspired, like self-supervised learning, on our way of learning.

The main concepts considered in transfer learning are the ones of *domain* and *action*:

- A *domain* $\mathcal{D}$, consists of a feature space $\mathcal{X}$ and a marginal probability defined on it;
- A *task* $\mathcal{T}$, is composed of a label space $\mathcal{Y}$ and a decision function defined as a conditional probability: $P(y_k|\mathbf{x}_j)$, $y_k \in \mathcal{Y}$, $k = 1, 2, ..., |\mathcal{Y}|$.

Transfer learning uses the knowledge in the source domain (or domains) and task (or tasks), here assumed a single domain and task, $\{\mathcal{D}_s, \mathcal{T}_s\}$, to improve the learned decision function in the target domain, $\{\mathcal{D}_t, \mathcal{T}_t\}$.

The transfer varies depending on the fact that $\mathcal{X}$, $\mathcal{Y}$ and the probability varies from the source to the target. The interested reader can refer to the surveys [4–6] and the numerous references therein.

18.4 Reinforcement Learning, RL

In RL an agent learns how to achieve a given objective [7, 8]. These agents are penalized when their decisions lead to a poor fulfillment of their objective and rewarded when they achieve good results. This constitutes precisely the reinforcement.

The main components of RL are: (i) the agent that takes actions; (ii) the possible actions; (iii) the so-called discount factor to makes immediate rewards to be more important than future rewards; (iv) the environement in which the agent operates; (v) the state that characterizes the situation in which the agent operates; (vi) the reward; (vii) the policy describing the agent strategy; (viii) the expected long-term return from a given state; and (ix) the so-called Q-value, similar to the previous one but that considers the action as an extra-parameter.

Reinforced learning aims at learning the policy π, that maps the space of states $\mathcal{S}$ and actions $\mathcal{A}$ into the unit interval, that is, $\pi : \mathcal{S} \times \mathcal{A} \to [0, 1]$, representing the probability of taking at the present time the action a when the state is given by s, i.e. $\pi(a, s) = P(a_t = a|s_t = s)$, to maximize the expected accumulated reward. Despite this probabilistic definition, deterministic policies also exist. In the same way, for a given state s, policy π and action a at time t_n, the updated state at time t_{n+1} could be either s', or the probability $P(s'|s, a, \pi)$.

The value associated with a policy, at a given state s, $V^\pi(s)$ reads

$$V^\pi(s) = \mathbb{E}\{R|\pi, s\} = \mathbb{E}\left\{\sum_{n=0}^{\infty} \gamma^n r(t_n)|s_0 = s, \pi\right\},$$

where $\mathbb{E}$ is the statistical expectation, R is the return, $r(t_n)$ the reward at time t_n and $\gamma \in [0, 1)$ is the discount rate, that being strictly lower than 1 implies that future are weighted less.

18.4 Reinforcement Learning, RL

The algorithm looks for the policy with maximum expected return. Usually, instead of making use of $V^\pi(s)$ (state-value), the so-called action-value $Q^\pi(s, a)$ is preferred, representing the expected cumulative reward from taking action a in state s, by following the policy

$$Q^\pi(s, a) = \mathbb{E}\{R|\pi, s, a\} = \mathbb{E}\left\{\sum_{n=0}^{\infty} \gamma^n r(t_n) | s_0 = s, a_0 = a, \pi\right\}.$$

The Bellman equation states that the optimal Q, noted Q^*, represents the maximum expected cumulative reward achievable from a state-action pair, i.e.,

$$Q^*(s, a) = \max_{\pi} \mathbb{E}\left\{\sum_{n=0}^{\infty} \gamma^n r(t_n) | s_0 = s, a_0 = a, \pi\right\}.$$

The optimal Q^* satisfies the so-called *Bellman equation*

$$Q^*(s, a) = \mathbb{E}_{s'}\left\{r + \gamma \max_{a'} Q^*(s', a')\right\},$$

where state-action values for the next time-step are referred by $\bullet'$.

The Bellman equation is solved by iterating from

$$Q_i(s, a) = \mathbb{E}_{s'}\left\{r + \gamma \max_{a'} Q_{i-1}(s', a')\right\},$$

iteration that converges when $i \to \infty$. That procedure is not scalable, becoming computationally infeasible.

The solution consists of using an approximation to estimate $Q(s, a)$, using for example a trained NN.

When considering Neural Networks in reinforcement learning, the NN is trained to link the state in the input layer with the Q-value for the possible actions (as many nodes in the output layer as number of possible actions). A valuable loss function considered in the back-propagation to calculate the NN parameters reads

$$\texttt{loss} = \text{Bellman}_{i-1} - Q(s, a; \theta_i),$$

with θ_i the network parameters, with

$$\text{Bellman}_{i-1} = \mathbb{E}_{s'}\left\{r + \gamma \max_{a'} Q^*(s', a'; \theta_{i-1})\right\}.$$

References

1. Y. LeCun, Self Supervised Learning. https://www.youtube.com/watch?v=SaJL4SLfrcY
2. S. Gidaris, P. Singh, N. Komodakis, Unsupervised representation learning by predicting image rotations, in *ICLR* (2018)
3. X. Zhu, *Semi-Supervised Learning Literature Survey*. University of Wisconsin-Madison
4. K. Weiss, T.M. Khoshgoftaar, D.D. Wang, A survey of transfer learning. J. Big Data **3**(9) (2016)
5. F. Zhuang, Z. Qi, K. Duan, D. Xi, Y. Zhu, H. Zhu, H. Xiong, Q. He, A comprehensive survey on transfer learning. Proc. IEEE **109**, 43–76 (2021)
6. S.J. Pan, J.T. Kwok, Q. Yang, Transfer learning via dimensionality reduction, in *Proceedings of the Twenty-Third AAAI Conference on Artificial Intelligence* (2008)
7. L.P. Kaelbling, M.L. Littman, A.W. Moore, Reinforcement learning: a survey. Artif. Intell. Res. **4**, 237–285 (1996)
8. M. van Otterlo, M. Wiering, Reinforcement learning and markov decision processes, in *Reinforcement Learning. Adaptation, Learning, and Optimization* ed. by M. Wiering, M. van Otterlo, vol. 12 (Springer, Berlin, Heidelberg, 2012)

Open Access This chapter is licensed under the terms of the Creative Commons Attribution-NonCommercial-NoDerivatives 4.0 International License (http://creativecommons.org/licenses/by-nc-nd/4.0/), which permits any noncommercial use, sharing, distribution and reproduction in any medium or format, as long as you give appropriate credit to the original author(s) and the source, provide a link to the Creative Commons license and indicate if you modified the licensed material. You do not have permission under this license to share adapted material derived from this chapter or parts of it.

The images or other third party material in this chapter are included in the chapter's Creative Commons license, unless indicated otherwise in a credit line to the material. If material is not included in the chapter's Creative Commons license and your intended use is not permitted by statutory regulation or exceeds the permitted use, you will need to obtain permission directly from the copyright holder.

Chapter 19
Regression: Basics

The present section revisits some usual regression techniques, widely considered in the next sections.

19.1 Polynomial Regression

In general—at least in the most basic setting—regression assumes that the quantity of interest (output) of our learning process, y, depends on the inputs $x_k, k = 1, \ldots, P$ through a polynomial law [1, 2].

The simplest choice is to assume that this dependence is linear with respect to the features x_i, i.e. $y(\mathbf{x}) = \beta_0 + \beta_1 x_1 + \cdots + \beta_P x_P$, where the $P + 1$ coefficients β_k are chosen to provide the best fit to the available data. If $M = 1 + P$ data are available, $\{x_1^{(j)}, \ldots, x_P^{(j)}, y^{(j)}\}, j = 1, \ldots, 1 + P$, we can write the matrix system

$$\begin{pmatrix} y^{(1)} \\ y^{(2)} \\ \vdots \\ y^{(P+1)} \end{pmatrix} = \begin{pmatrix} 1 & x_1^{(1)} & x_2^{(1)} & \cdots & x_P^{(1)} \\ 1 & x_1^{(2)} & x_2^{(2)} & \cdots & x_P^{(2)} \\ \vdots & \vdots & \vdots & \ddots & \vdots \\ 1 & x_1^{(P+1)} & x_2^{(P+1)} & \cdots & x_P^{(P+1)} \end{pmatrix} \begin{pmatrix} \beta_0 \\ \beta_1 \\ \vdots \\ \beta_P \end{pmatrix},$$

whose solution will provide us with the β_k coefficients, resulting in the linear regression $y(\mathbf{x})$.

When the number of available data M is smaller or larger than $P + 1$, the previous system of equations results under or over-determined, respectively. To overcome this limitation, different techniques exist: pseudo-inverses, L2- or L1-optimization, to name but a few. The reader can readily establish a connection to compressed sensing, see Sect. 10, or the usual Matlab™ backslash.

Linear regression making use of linear approximation constitutes an appealing choice due to its limited need of data, always of the same order than the number of the features involved in the approximation. However, as it is frequently the case in many applications, linear dependencies do not always explain the observed phenomena, and non-linear approximations into linear regressions should be preferred.

Extending linear regression to non-linear approximations, by considering a higher-degree polynomial, does not pose major difficulties if the number of involved features remains limited. To develop a quadratic approximation, we must consider

$$y = \beta_0 + \sum_{i=1}^{P} \beta_i x_i + \sum_{i=1}^{P} \sum_{j \geq i}^{P} \beta_{ij} x_i x_j,$$

where the number of coefficients (and consequently the required amount of data) scales with P^D, the sought the approximation degree.

Thus, if the number of dimensions in which data lives, P, remains reasonably small, the approximation degree D could be increased without major difficulties. In the high-dimensional, multi-parametric case, however, first-degree polynomial linear regression should be preferred to avoid the so-called curse of dimensionality. The interest on developing multi-parametric regressions compatible with high-degree approximations and a limited amount of sampled data is obvious. This topic will be addressed later.

19.2 Kriging

In fact Kriging was already introduced when addressing Gaussian processes in Sect. 14. Here, we are giving an alternative derivation that fits better to the usual description given in the literature [3]. For this purpose we will derive the linear estimation formulation.

We consider an inference problem, that looks for y at an unobserved point $\mathbf{x}$, from M observed quantities in the neighborhood of $\mathbf{x}$, $\mathbf{x}_i$, $i = 1, \ldots, M$, with values $y_1, y_2, \ldots, y_M$.

Thus, the linear estimator writes

$$\hat{y}(\mathbf{x}) = \sum_{i=1}^{M} \omega_i(\mathbf{x}) y(\mathbf{x}_i) = \mathbf{W}^T \mathbf{Y},$$

where the weights ω_i should guarantee: (i) unbiased estimation; and (ii) minimum variance. We are developing both points.

We define the error $\epsilon(\mathbf{x})$ as the difference between the linear estimation $\hat{y}(\mathbf{x})$ and the real value $y(\mathbf{x})$

$$\epsilon(\mathbf{x}) = \sum_{i=1}^{M} \omega_i(\mathbf{x}) y(\mathbf{x}_i) - y(\mathbf{x}) = \hat{y}(\mathbf{x}) - y(\mathbf{x}) = \tilde{\mathbf{W}}^T \tilde{\mathbf{Y}},$$

with $\tilde{\mathbf{W}}^T = (\mathbf{W}^T, -1)$ and $\tilde{\mathbf{Y}}(\mathbf{x})^T = (y(\mathbf{x}_1), y(\mathbf{x}_2), \ldots, y(\mathbf{x}_M), y(\mathbf{x})) = (\mathbf{Y}^T, y(x))$.

Being the random function stationary $\mathbb{E}(y(\mathbf{x}_i)) = \mathbb{E}(y(\mathbf{x})) = \mu$, from which the lack of bias $\mathbb{E}(\epsilon(\mathbf{x})) = 0$ implies

$$\mathbb{E}(\epsilon(\mathbf{x})) = 0 \rightarrow \mu \sum_{i=1}^{M} \omega_i(\mathbf{x}) - \mu = 0,$$

or

$$\sum_{i=1}^{M} \omega_i(\mathbf{x}) = \mathbf{I}^T \mathbf{W} = 1.$$

Now, we are enforcing the minimum variance (to guarantee smoothness). The variance of $\epsilon(\mathbf{x})$, with $\mathbb{E}(\epsilon(\mathbf{x})) = 0$, reads

$$\mathrm{Var}(\epsilon(\mathbf{x})) = \tilde{\mathbf{W}}^T \, \mathrm{Cov} \, \tilde{\mathbf{W}},$$

with matrix Cov containing the covariances of the random variables $(y(\mathbf{x}_i), y(\mathbf{x}_j))$, $(y(\mathbf{x}_i), y(\mathbf{x}))$ and $(y(\mathbf{x}), y(\mathbf{x}))$.

In a more developed form the variance can be written as

$$\mathrm{Var}(\epsilon(\mathbf{x})) = \mathrm{Cov}(y(\mathbf{x}), y(\mathbf{x})) + \sum_{i=1}^{M}\sum_{j=1}^{M} \omega_i \omega_j \, \mathrm{Cov}(y(\mathbf{x}_i), y(\mathbf{x}_j)) - 2\sum_{i=1}^{M} \omega_i \, \mathrm{Cov}(y(\mathbf{x}_i), y(\mathbf{x})),$$

that must be minimized with the constraint of lack of bias $\sum_{i=1}^{M} \omega_i = 1$, added as a Lagrange multiplier before enforcing the minimization that will finally lead to the searched coefficients $\omega_i(\mathbf{x})$, as well as the Lagrange multiplier.

19.3 Support Vector Regression, SVR

Support Vector Regression, SVR, [4], has some ingredients in common with Support Vector Machines. Remember its use in supervised classification, already addressed in Sect. 16.

SVR regression begins by considering $y = \beta_0 + \mathbf{W}^T \mathbf{x}$, and functional $\mathcal{G}(\mathbf{W})$

$$\mathcal{G}(\mathbf{W}) = \frac{1}{2} \mathbf{W}^T \mathbf{W},$$

that is minimized while enforcing as constraint a regularized form of the loss $|y_s - \beta_0 - \mathbf{W}^T \mathbf{x}_s| \leq \epsilon$, $s = 1, \ldots, M$, leading to

$$\mathcal{G}(\mathbf{W}) = \frac{1}{2} \mathbf{W}^T \mathbf{W} + C \sum_{s=1}^{M} \xi_s \xi_s^*,$$

with $\xi_s \geq 0$ and $\xi_s^* \geq 0$,

$$y_s - (\mathbf{W}^T \mathbf{x}_s + \beta_0) \leq \epsilon + \xi_s,$$

and

$$(\mathbf{W}^T \mathbf{x}_s + \beta_0) - y_s \leq \epsilon + \xi_s^*.$$

Of course, this is not the only possibility. Many other alternatives to address the nonlinear case exist.

19.4 Likelihood-Based Regressions for Noisy Data

The maximum likelihood method allows efficient estimation of the parameters involved in regression models, when available data exhibit a non negligible noise.

Whereas least-squares try to minimize the sum of the squared errors, the likelihood method looks for the parameters having a large probability to describe the sample data.

For describing it, we consider the simplest linear regression

$$y_t = a_0 + a_1 x_t + u_t,$$

with y_t the measure al location x_t, a_0 and a_1 the regression coefficients and u_t the noise, normally distributed, i.e. $u_t \sim \mathcal{N}(0, \sigma_u^2)$.

Thus, the statistical expectation results $\mathbb{E}(y_t) = a_0 + a_1 x_t$ and the variance of y_t, $\mathbb{V}(y_t) = \sigma^2$, from which, it results $y_t \sim \mathcal{N}(a_0 + a_1 x_t, \sigma_u^2)$.

For t measures, the joint probability reads:

$$f(y_1, \ldots, y_t | a_0 + a_1 x, \sigma_u^2),$$

that assuming y_k, $k = 1, \ldots, t$, independent, can be rewritten as

$$f(y_1, \ldots, y_t | a_0 + a_1 x, \sigma_u^2) = f(y_1 | a_0 + a_1 x_1, \sigma_u^2) \cdot f(y_2 | a_0 + a_1 x_2, \sigma_u^2) \cdot \ldots \cdot f(y_t | a_0 + a_1 x_t, \sigma_u^2),$$

with each one expressible from

$$f(y_k | a_0 + a_1 x_k, \sigma_u^2) = \frac{1}{\sqrt{2\pi}\sigma_u} e^{-\frac{1}{2}\left(\frac{y_k - a_0 - a_1 x_k}{\sigma_u}\right)^2},$$

from which the joint distribution becomes

$$f(y_1, \ldots, y_t | a_0 + a_1 x, \sigma_u^2) = \frac{1}{(\sqrt{2\pi})^t \sigma_u^t} e^{-\frac{1}{2} \sum_{n=1}^{t} \left(\frac{y_n - a_0 - a_1 x_n}{\sigma_u}\right)^2}.$$

19.4 Likelihood-Based Regressions for Noisy Data

The right-hand member logarithm reads

$$-\frac{t}{2}\ln(2\pi) - t\ln(\sigma_u) - \frac{1}{2}\sum_{n=1}^{t}\left(\frac{y_n - a_0 - a_1 x_n}{\sigma_u}\right)^2,$$

that taking into account $\frac{1}{2}\ln(\sigma_u^2) = \ln(\sigma_u)$, can be rewritten as

$$-\frac{t}{2}\ln(2\pi) - \frac{1}{2}\ln(\sigma_u^2) - \frac{1}{2}\sum_{n=1}^{t}\left(\frac{y_n - a_0 - a_1 x_n}{\sigma_u}\right)^2. \tag{19.1}$$

Now, for estimating the parameters, it suffices taking the derivatives of Eq. (19.1) with respect to a_0, a_1 and σ_u and enforcing all them to vanish:

- $\partial/\partial a_0 = 0$ implies:

$$\sum_{n=1}^{t}\frac{y_n - a_0 - a_1 x_n}{\sigma_u} = 0,$$

or

$$\sum_{n=1}^{t} y_n = t \cdot a_0 + a_1 \sum_{i=1}^{t} x_n.$$

- $\partial/\partial a_1 = 0$ implies:

$$\sum_{n=1}^{t}\frac{y_n - a_0 - a_1 x_n}{\sigma_u} x_n = 0,$$

or

$$\sum_{n=1}^{t} y_n x_n = a_0 \sum_{i=1}^{t} x_n + a_1 \sum_{i=1}^{t} x_n^2.$$

- $\partial/\partial \sigma_u^2 = 0$ implies, after some routine manipulations [5]:

$$-t + \frac{1}{\sigma_u^2}\sum_{n=1}^{t}(y_n - a_0 - a_1 x_n)^2 = 0,$$

or

$$\sigma_u^2 = \frac{1}{t}\sum_{n=1}^{t}(y_n - a_0 - a_1 x_n)^2 = 0.$$

The first two equations coincide with the ones obtained by using least-squares, however the estimator for the variance differs, being the just obtained biased. The unbiased least-squares estimator reads:

$$\sigma_u^2|_{\text{unbiased}} = \frac{1}{t-2} \sum_{n=1}^{t} (y_n - a_0 - a_1 x_n)^2 = 0.$$

The just described procedure can be combined with any higher degree polynomial regression, looking for the coefficients that allow distributing the data around the response surface in order to fulfill the data variability (when known) or minimizing the variance without an *a priori* knowledge on the data variability.

References

1. J. Fan, Local polynomial modelling and its applications: from linear regression to nonlinear regression, in *Monographs on Statistics and Applied Probability* (Chapman & Hall/CRC, 1996)
2. N. Kutz, *Data-Driven Modeling & Scientific Computation: Methods for Complex Systems & Big Data* (Oxford University Press, Oxford, 2013)
3. A. Papritz, A. Stein, Spatial prediction by linear kriging, in *Spatial Statistics for Remote Sensing. Remote Sensing and Digital Image Processing* ed. by A. Stein , F. Van der Meer, B. Gorte, vol. 1 (Springer, Dordrecht, 1999)
4. M. Awad, R. Khanna, Support vector regression, in *Efficient Learning Machines* (Apress, Berkeley, CA, 2015)
5. J. Kibala Kuma, Estimation par la methode du Maximum de Vraisemblance (2019). https://hal.archives-ouvertes.fr/cel-02189969

Open Access This chapter is licensed under the terms of the Creative Commons Attribution-NonCommercial-NoDerivatives 4.0 International License (http://creativecommons.org/licenses/by-nc-nd/4.0/), which permits any noncommercial use, sharing, distribution and reproduction in any medium or format, as long as you give appropriate credit to the original author(s) and the source, provide a link to the Creative Commons license and indicate if you modified the licensed material. You do not have permission under this license to share adapted material derived from this chapter or parts of it.

The images or other third party material in this chapter are included in the chapter's Creative Commons license, unless indicated otherwise in a credit line to the material. If material is not included in the chapter's Creative Commons license and your intended use is not permitted by statutory regulation or exceeds the permitted use, you will need to obtain permission directly from the copyright holder.

Chapter 20
Neural Network Based Machine Learning Techniques

A valuable introduction to neural networks can be found in [1], with numerous variants revisited in the next sections, enabling the construction of nonlinear regressions operating on different typologies of data (lists, images, graphs, time series, ...).

20.1 From the Neuron to Deep Neural Networks

Deep-learning is based on the use of neural networks, NN, with several layers (hence the adjective "deep"). Roughly speaking, NNs take some input data and generate an output, after transiting throughout the different neurons of the different layers.

To describe NNs in more detail, take a look to the sketch in Fig. 20.1. This simple (the simplest, in fact) artificial neuron receives two input data, x_1 and x_2, and produces an output y. The simplest option is to produce a linear combination of the inputs as output, by multiplying each one by a weight, W_1 and W_2: $y = W_1 x_1 + W_2 x_2$. In a more general scenario, many inputs could exist (think of P different ones) so that

$$y = \sum_{i=1}^{P} W_i x_i,$$

or, in other words,

$$y = \mathbf{W}^T \mathbf{x}. \tag{20.1}$$

Equation (20.1) states that the output is a linear combination of the input data. Very often, again, this linear assumption will not be enough, but this will be addressed later on.

Before using Eq. (20.1), the best vector $\mathbf{W}$ that explains the input-output relationship must be computed. If an input-output couple is available $(\mathbf{x}^{(1)}, y^{(1)})$, (we assume

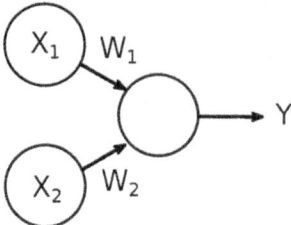

Fig. 20.1 Sketch of a simple neuron

a normalized input, i.e. $\|\mathbf{x}^{(1)}\| = 1$), the sought vector will be $\mathbf{W} = y\mathbf{x}^{(1)}$. This ensures the recovery of the known output at the input point. So far, no learning has been produced, since we still do not know anything about previously unseen input points, different to those of the training procedure (the phase in which we determine the best $\mathbf{W}$).

In a general setting, more than a single input-output couple, M couples $(\mathbf{x}^{(i)}, y^{(i)})$, $i = 1, \ldots, M$, will be available. If the inputs are orthonormal $\mathbf{x}^{(i)} \cdot \mathbf{x}^{(j)} = \delta_{ij}$ (with δ the Kronecker delta), then it is easy to prove that

$$\mathbf{W} = \sum_{i=1}^{M} y^{(i)} \mathbf{x}^{(i)}.$$

In the more general case, the problem of finding the best weighting vector results in a least-squares minimization problem. Thus, a functional depending on the weights, $\epsilon(\mathbf{W})$, can be defined

$$\epsilon(\mathbf{W}) = \sum_{i=1}^{M} \left(y^{(i)} - \mathbf{W}^T \mathbf{x}^{(i)} \right)^2,$$

that, by defining $\mathbf{Y}^T = \left(y^{(1)} \; y^{(2)} \; \cdots \; y^{(M)} \right)$ and $\mathbf{X} = \left(\mathbf{x}^{(1)} \; \mathbf{x}^{(2)} \; \cdots \; \mathbf{x}^{(M)} \right)$, the previous expression can be rewritten as

$$\epsilon(\mathbf{W}) = \frac{1}{2} \left(\mathbf{Y}^T - \mathbf{W}^T \mathbf{X} \right)^2,$$

whose minimization reads $\mathbf{X}\mathbf{X}^T \mathbf{W} = \mathbf{X}\mathbf{Y}$.

For alleviating the computational complexity, different iterative algorithms have been proposed, most of them based on a gradient minimization of the functional $\epsilon(\mathbf{W})$.

Until now, everything remains linear. For this reason the performance of the proposed learning procedure remains quite limited. Thus, a key ingredient has been added to the previous methodology that consists in including a nonlinear function $\sigma(\cdot)$ that determines the activation of the neuron. Different choices exist for that function involved in the functional

$$\epsilon(\mathbf{W}) = \frac{1}{2} \left(\mathbf{Y}^T - \sigma\left(\mathbf{W}^T \mathbf{X} \right) \right)^2,$$

that combined with iterative adaptive procedures for minimizing it, allows the calculation of **W**.

When P-dimensional inputs produce multidimensional outputs, that is, for each $\mathbf{x} \in \mathbb{R}^P$ corresponds a vector output $\mathbf{y} \in \mathbb{R}^Q$, it suffices considering Q neurons, each one having as input **x** and having as output the component q of **y**, with $q = 1, \ldots, Q$.

In this case **W** becomes a matrix instead of a vector. Assuming that M couples $(\mathbf{x}^{(i)}, \mathbf{y}^{(i)})$, $i = 1, \ldots, M$, are known, the computation of this matrix is very similar to the one already described.

Very often, instead of considering a single layer of neurons, it has been noticed that using multiple layers of neurons give better results. In this way, the output of a layer becomes the input of the next, and so on, until reaching the output.

The parameters defining the NN architecture and functioning (e.g. number of layers, number of neurons in each layer, activation function, ...) are called hyper-parameters, while the neuron weights are called parameters. (Hyper-)parameters are determined by minimizing the so-called *loss* function that represents the difference (in some appropriate norm) between the known experimental data and the output produced by the NNs.

20.2 The Universal Approximation Theorem

The performance of Neural Networks follows the *universal approximation theorem*. This theorem states that any nonlinear function can be approximated very accurately from a single layer of neurons, under reasonable assumptions, as the non-polynomial form of the activation function.

A valuable illustration of that property is given in [2], where the authors provide a visual construction. For that purpose, that reference considers the data x communicated to the layer composed of N neurons, as depicted in Fig. 20.2, where the n-neuron computes $\omega_n x + b_n$, before applying the activation function $\sigma(\cdot)$, for example the usual sigmoid, $\sigma(\xi) = 1/(1 + e^{-\xi})$.

All the contributions are collected in the output layer, with a single output that performs the weighted sum of all the values coming from all the neurons of the

Fig. 20.2 Illustrating the universal approximation theorem

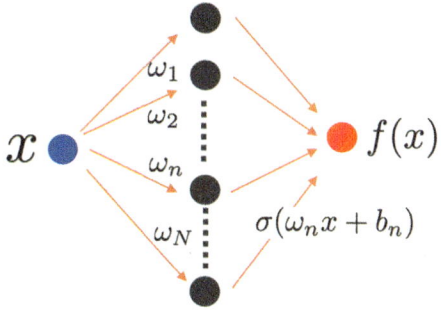

hidden layer, i.e.

$$\sum_{n=1}^{N} h_n \, \sigma(\omega_n x + b_n) \approx f(x).$$

Considering a generic neuron in the hidden layer (avoiding the use of index $\bullet_n$ for alleviating the notation), it performs according to $\sigma(\omega x + b)$.

It can be noticed that by changing b, the resulting function $\sigma(\omega x + b)$ translates on the x-axis, and that by increasing the value of the weight ω, function $\sigma(\omega x + b)$ approaches the step function:

$$\sigma(\omega x + b) \approx \begin{cases} 0 \text{ if } x < s \\ 1 \text{ if } x \geq s \end{cases},$$

with $s = -b/\omega$ and ω large enough. Thus, the weight to bias ratio, b/ω, determines the location of the *quasi-step* function.

In summary, it suffices considering a number large enough of neurons, with the weight to bias ratio of each neuron chosen for distributing their associated *steps* along the x−axis, such that by applying the adequate weight h_n to each step function, the target function can be approximated.

Thus, we can conclude that NN can approximate general nonlinear functions, from an adequate choice of the number of neurons, the number of layers, and the different weights and bias ratios.

20.3 Learning Operators: *DeepONets*

The universal approximation theorem that represents one of the major achievements at the origin of the numerous NN successes, was generalized to functionals and operators [3, 4]. For the sake of completeness, both results are given below (see the Appendix of [5] for additional details):

- *Universal approximation theorem for functionals.*

For any functional $f(u)$, with $u(x)$ in an appropriate functional space, the universal theorem, when considering m points, reads

$$\left| f(u) - \sum_{i=1}^{n} c_i \sigma \left(\sum_{j=1}^{m} \xi_{ij} u(x_j) + \theta_i \right) \right| < \epsilon,$$

with $\epsilon > 0$ small enough, n a positive integer, and c_i, θ_i and ξ_{ij} real constants.
- *Universal approximation theorem for operators.*

For an operator G, that applied on $u(x)$ produces $G(u)$, its evaluation at y reads $G(u)(y)$. The universal theorem reads in the present case

$$\left| G(u)(y) - \sum_{k=1}^{p}\sum_{i=1}^{n} c_i^k \sigma\left(\sum_{j=1}^{m} \xi_{ij}^k u(x_j) + \theta_i^k\right) \sigma(\omega_k \cdot y + \varphi_k) \right| < \epsilon, \quad (20.2)$$

with $\epsilon > 0$ small enough, p and n positive integers, and c_i^k, θ_i^k, ξ_{ij}^k and φ_k real constants, and $\omega_k \in \mathbb{R}^d$.

The result stated in Eq. (20.2) was employed in [5] to address operators approximation and more specifically for learning them from data. As this expression indicates, the input data consists of the m values $u(x_j)$, $j = 1, ..., m$, as well as the value of y. The m values enter into p nets, whose respective outputs are: $\sum_{i=1}^{n} c_i^k \sigma\left(\sum_{j=1}^{m} \xi_{ij}^k u(x_j) + \theta_i^k\right) \equiv b_k$, $k = 1, ..., p$. On the other hand, the last network takes y as input and computes $\sigma(\omega_k \cdot y + \varphi_k) \equiv t_k$. Then, it suffices calculating the sum $\sum_{k=1}^{p} b_k t_k$ to obtain the desired output, i.e. $G(u)(y)$.

20.4 Convolutional NN, CNN (ConvNet)

CNN are widely applied in computer vision [6]. The CNN input consists of a tensor whose size is the number of images times the image height times the image width times the image depth (the channels).

The NN computes its parameters (weights and biases) to extract features in the image enabling to differentiate them.

CNN avoids the need of flattening the tensor (that is, to convert the tensor into a vector as usual NN perform) and consequently it conserves the spatial and temporal dependencies.

Because of the images dimension, they are reduced prior to process. For that purpose different units compose the CNN, as summarized below:

- Convolution layers employ a kernel, also called filter, learned during the CNN training.

We consider for simplicity the kernel depth equal to one, and consequently the kernel becomes a small-size matrix that moves on the image (big-size matrix). At each position on it, the different components of the filter multiply the corresponding ones in the matrix. Then, the convolution at the present position of the kernel is simply the sum of all the just referred products.

As indicated, the kernel moves on the image (the step with respect to the image pixels is called stride), until it visists the whole image. The coverage has different variants, referred as padding.

For the sake of simplicity, we consider a convolution unit (layer) that proceeds with a unit stride and a conventional padding, acting on an image **I** of size $N \times N$ (same height and width) whose components are noted by $I(i, j)$, and the kernel **K** of size $M \times M$ (again same height and width), with $M \ll N$, being its components noted by $K(r, s)$. The convolution results are grouped in matrix **C** of size $P \times P$, with $P = N - M + 1$ (for the considered stride and padding).

At each position (m, n) of the kernel on the input image, the convolution reads

$$C(m, n) = \sum_{i=1}^{M} \sum_{j=1}^{M} K(i, j) \cdot I(m + i - 1, n + j - 1).$$

Convolution extracts low-order features (e.g. edges) and with added layers, higher-level features as well.
- The so-called *pooling layer* reduces the spatial size of the convolved feature. It aims at reducing the processing computational cost as well as to attenuate the effect of noise.

Among the several pooling strategies, the so-called *Max Pooling* returns the maximum value within the region of the image covered by the kernel support. Equivalently, *Average Pooling* returns the average of all the values in this same region.
- At the end a fully connected layer, acting on the flattened reduced (convolved and pooled) image, is used for learning purposes.

There are many variants of CNNs, among them Autoencoders, UNet, DenseNet, YOLO, ... to name few.

20.5 Graph Neural Network, GNN, and Message Passing, MP

One of the main issues of usual NN is that they proceed from flattened data. Thus, if one considers a finite element mesh, and considers nodal variables as the data for learning a model relating nodal values representing the input, with the associated nodal values representing the output, those vectors do not contain the connectivity information, that within a finite element framework is retained in the stiffness matrix. Thus, NN must learn that connectivity from the data during the training process, with the associated cost.

Convolutional NN alleviate such an issue, by using, as just described, convolution and pooling layers. Convolution operates easily in regular grids, as for example images, however, convolution must be redefined in the case of unstructured grids, that is, general graphs.

20.5 Graph Neural Network, GNN, and Message Passing, MP

Graph Neural Networks aim at retaining connectivities, that express the real physical links, in general, distance dependent. Thus, in a certain sense, one is aiming at discovering the model describing the evolution of nodal values, as well as the one related to the edge features. From a physical point of view, in some cases, the former accounts for conservation, while the later is expected describing fluxes.

Some GNN look for the model at the graph level, and takes into account different graph topologies, sizes (number of vertices), ... by using the so-called *zero padding* or a *disjoint union matrix*.

Other possibility, the so-called *message passing*, operates at the level of vertices and edges, the later always involving two vertices, the two defining the edge. The invariance with respect to permutations, solves the apparent problem of having vertices connected to different number of edges. In fact, the vertex model depends for example in the sum of all the information coming from all the edges connected to the node (ensuring permutation invariance) and that reduces to a single value. This fact allows training a unique vertex model independently to the number of edges connected with the vertex.

Should the reader be interested in GNN, we recommend the reading of [7] and the numerous references therein.

Let $\mathcal{G} = (\mathcal{V}; \mathcal{E}; u)$ be a directed graph, where $\mathcal{V}$ the set of vertices, $\mathcal{E}$ the set of edges, and u the set of global features.

It is common practice to assume the equivalence of each vertex and edge in the graph with the (finite element, finite difference, ...) nodes in a discretized physical system and their pairwise interaction between nodes, respectively.

In turn, the global feature vector, groups properties or parameters shared by the whole graph, like gravity or material properties. Finally, a feature vector v_i is associated to each vertex, representing the physical properties of each individual node in the physical model. Similarly, to each edge joining nodes i and j, a edge feature e_{ij} is attached.

Before looking for the vertex and edge models, an encoding and decoding units are integrated at both levels, the one of the vertex and the one of the edge, to increase the data dimensionality before learning the models. Thus, v_i is mapped into x_i, and e_{ij} into x_{ij}.

Then, the processing unit establishes both, the vertex and the edge models bu using multi-layer perceptron units, MLP, and message passing, MP:

- The edge-MLP model reads:

$$(x_{ij}, x_i, x_j, u) \to x'_{ij}.$$

- The vertex-MLP proceeds at its turns from:

$$(x_i, \phi(x'_{ij}), u) \to x'_i,$$

where $\phi(x'_{ij})$ combines all the contributions arriving at the i-th vertex from any edge connected to it, and ensuring permutation invariance, for example, among the many possible choices:

$$\phi(x'_{ij}) = \sum_{j \in \mathcal{S}_i} x'_{ij},$$

with $\mathcal{S}_i$, the set of vertices connected to vertex i from any edge in the set $\mathcal{E}$.

Thus, the processing step just described represents the so-called *message passing*. To obtain the influence of graph nodes placed at distant positions, the message passing process can be recurrently repeated a number of times. The final non-local character of the resulting model depends on how many of these passings are accomplished.

In [8] the model with a graph support was learnt while ensuring thermodynamic consistency, a natural extension of the structure preserving neural networks introduced later.

20.6 Recurrent Neural Networks, rNN

The objective of Machine Learning is inferring the output related to a given input. For instance, in a structural dynamics problem the input could be the applied load (which eventually may evolve in time) and the output could be the measured displacement at some locations of the structure.

Machine Learning tries to unveil the relationship between these observables, the input action and the output response. However, in many cases such a direct one-to-one relation does not exist. Very often, the output depends on some internal variables of the system, e.g. the previous state.

Recurrent Neural Networks (rNN) address such a situation by including the time evolution of the internal state at the same time that it construct the model relating the observed input and output (action and response).

In [9], the authors addressed structural dynamics and created a NN for predicting the response at a given time step, from the internal state at the previous time step $t-1$, $\mathbf{h}_{t-1}$, and the loading at the present time, $\mathbf{x}_t$. From them, the trained model predicts the present internal state $\mathbf{h}_t$ and the output of interest $\mathbf{y}_t$.

If $\mathbf{W}^\bullet$ represents dense matrices associated to variables $\bullet$, $\sigma(\cdot)$ the activation function, $\mathbf{h}_t$ the internal state at time t, and $\mathbf{x}_t$ and $\mathbf{y}_t$ the associated input and output at the present time t, then the rNN proceeds from:

$$\begin{cases} \mathbf{h}_t = \sigma^h(\mathbf{W}^h \mathbf{h}_{t-1} + \mathbf{W}^x \mathbf{x}_t) \\ \mathbf{y}_t = \sigma^y(\mathbf{W}^y \mathbf{h}_t), \end{cases}$$

whose architecture is sketched in Fig. 20.3.

To perform time integration it suffices to use the rNN and re-inject the rNN output $\mathbf{h}_t$, as input at the next time step.

There are many possible rNN architectures and a vast literature addressing them.

Fig. 20.3 Recurrent Neural Network architecture

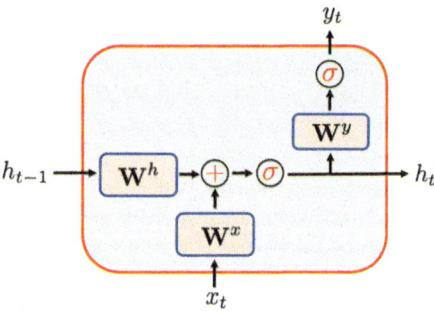

20.7 Long Short Time Memory, LSTM

As just commented, there are many possible architectures of rNN for addressing more complex scenarios. In particular the long short time memory, LSTM, that allows longer memory effects than the rNN just described.

LSTM combines a short and a long time memory units, with an evanescent memory for the long-time path, and a combination of long and short leads to the short memory response, [10–12].

The LSTM structure, taken from [13], is illustrated Fig. 20.4 for a unit cell. The green line in that figure represents the long time memory path, the blue line the short time memory one, while red and orange lines represent respectively the input and the output paths.

The single unit proceeds according to:

$$\begin{cases} \mathbf{f}_t = \sigma\left(\mathbf{W}_f\left[\mathbf{h}_{t-1}, \mathbf{x}_t\right] + \mathbf{b}_f\right) \\ \mathbf{i}_t = \sigma\left(\mathbf{W}_i\left[\mathbf{h}_{t-1}, \mathbf{x}_t\right] + \mathbf{b}_i\right) \\ \tilde{\mathbf{c}}_t = \tanh\left(\mathbf{W}_c\left[\mathbf{h}_{t-1}, \mathbf{x}_t\right] + \mathbf{b}_c\right) \\ \mathbf{o}_t = \sigma\left(\mathbf{W}_o\left[\mathbf{h}_{t-1}, \mathbf{x}_t\right] + \mathbf{b}_o\right) \\ \mathbf{c}_t = \mathbf{f}_t \times \mathbf{c}_{t-1} + \mathbf{i}_t \times \tilde{\mathbf{c}}_t \\ \mathbf{h}_t = \mathbf{o}_t \times \tanh(\mathbf{c}_t) \end{cases} \quad (20.3)$$

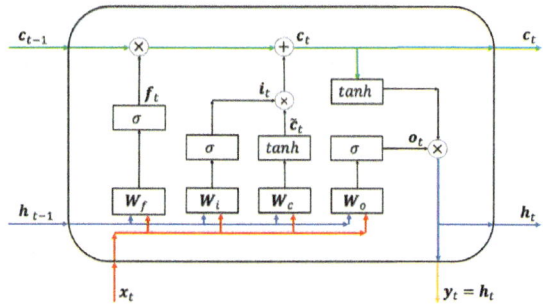

Fig. 20.4 A single LSTM unit ($\times$ refers to the Hadamard term-by-term multiplication, and symbol $+$ refers to the sum). tanh and σ refers respectively the hyperbolic tangent and sigmoid activation functions

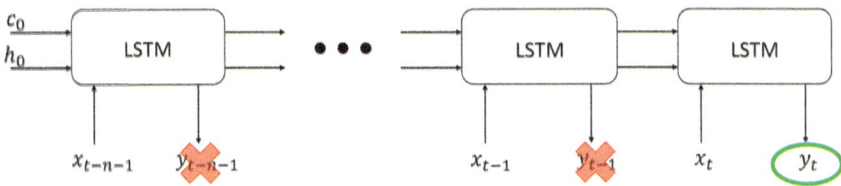

Fig. 20.5 LSTM-block architecture: The output y_t results from the previous n inputs while it almost totally forget the initial hidden long and short memory states h_0 and c_0 that can be initialized with zero values

where σ represents the sigmoid activation function and tanh the hyperbolic tangent. Equation (20.3) reproduces the forward propagation in a single LSTM unit. $\mathbf{W}_\bullet$ and $\mathbf{b}_\bullet$ correspond to the different matrices containing the network weights and biases to be learnt.

The LSTM unit illustrated in Fig. 20.4 contains a *forget gate* for the long time memory $\mathbf{c}_{t-1}$, that with $\mathbf{f}_t \in [0, 1]$, enables "forgetting" a part of the registered memory. The *forget* operation is followed by an update of the long time memory, using an *input gate* $\mathbf{i}_t$, acting on $\tilde{\mathbf{c}}_t$. Finally, an output gate $\mathbf{o}_t$ is applied to the long time memory to provide, in this case, the output $\mathbf{y}_t = \mathbf{h}_t$, while $\mathbf{h}_t$ is sent to the next LSTM unit through the short time memory path.

A particular architecture combines several LSTM (or rNN) units, for accounting for the n previous states, within the so-called LSTM-block (or rNN-block), as sketched in Fig. 20.5, that chains n LSTM units as the ones just described.

The LSTM-block unit (the so-called rNN-block operates similarly) proceeds from an arbitrary choice of the initial state of both, the short and long memory states, $\mathbf{h}_0$ and $\mathbf{c}_0$ respectively. After transiting through the n units, subjected to the injected inputs $\mathbf{x}_t, \cdots, \mathbf{x}_{t-n-1}$, the LSTM-block output $\mathbf{y}_t$ has almost totally forgot the initial hidden states $\mathbf{h}_0$ and $\mathbf{c}_0$, and its value is almost dictated by the n most recent inputs. Thus, only the output of the last LSTM unit composing the LSTM-block is used. This architecture allows using the LSTM-block as an integrator.

One could expect that the use of LSTM is restricted to models involving high order time derivatives, because first order time derivative only involves the state at the present and the previous time steps, a quite reduced memory that rNN is expected capturing without major difficulties. However, sometimes there are hidden (unobserved) dynamics coupled with the considered (observable) one, and the effect of these hidden dynamical variables can be captured from a larger time memory, as stated by the Taken delay embedding, as discussed in [14].

20.8 Gated Recurrent Unit, GRU

Looking for higher performances with respect to rNN, rNN were modified for accommodating a single gate (instead of the two characteristic of LSTM) while offering more flexibility on the internal state updating, with fewer parameters to train than LSTM.

There are many architectures: fully gated unit; minimal gated unit; light gated recurrent unit, ... to mention a few. The interested reader can refer to [15] for a evaluation of those units.

20.9 Reservoir Computing

Reservoir computing maps input data into higher dimensional spaces by using a fixed, non-linear system, non-trainable sparsely connected recurrent part, the so-called reservoir, whose weights are generated randomly and remain unchanged during the training.

When the data reach the reservoir, considered as a *black box*, a trainable layer reads the state of the reservoir and computes the output.

A valuable description of the functioning of the so-called Echo State Network (ESN) can be found in [16].

20.10 Residual Nets

Residual nets (ResNet) try to imitate the backward integration of a dynamical system, and instead of inferring the state updating (as rNN or LSTM perform), ResNet aims at learning the dynamical system model.

When considering the dynamical system

$$\frac{d\mathbf{x}}{dt} = \mathbf{f}(\mathbf{x}), \quad \mathbf{x}(t=0) = \mathbf{x}_0,$$

the ResNet looks for the solution at time t_{n+1} knowing the solution at time t_n, with $t_{n+1} - t_n = \Delta t$.

By considering the simplest time steeping, the time derivative can be approximated by $d\mathbf{x}/dt = (\mathbf{x}_{n+1} - \mathbf{x}_n)/\Delta t$, and then one can use the following updating

$$\mathbf{x}_{n+1} = \mathbf{x}_n + \Delta t \mathbf{f}(\mathbf{x}_n),$$

and learn the updating $\Delta t \mathbf{f}(\mathbf{x}_n) \equiv \tilde{\mathbf{f}}(\mathbf{x}_n)$.

For that purpose ResNet creates a NN for learning from $\mathbf{x}_{n+1} - \mathbf{x}_n$, the forcing function $\tilde{\mathbf{f}}(\mathbf{x})$. From the trained NN, the output $\mathbf{x}_{n+1}$ is calculated by applying on $\mathbf{x}_n$ the NN while adding the identity operator applied on the input.

Now, the dynamical system integration proceeds by applying in a recursive way the ResNet, with the output at time n being the input of the next time-step. The ResNet architecture and the integration procedure are illustrated in Fig. 20.6.

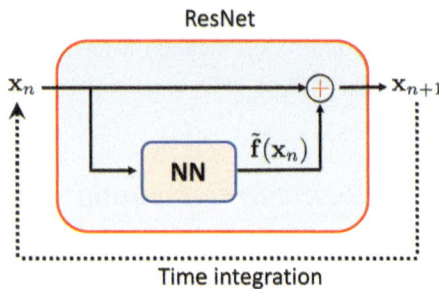

Fig. 20.6 ResNet global architecture and integration mode (dotted line)

20.11 NeuralODE

In the ResNet just described, the learning only involves two consecutive time steps, the ones involved by the state at present time, and the state at the previous time step.

Even if the ResNet training procedure remains conceptually correct, numerical errors can motivate that the learnt model fails to ensure time integration stability, that is, when using the just learnt model to integrate in time the dynamical system. This issue was deeply discussed in [14], where some strategies for ensuring stability were proposed.

One possibility for alleviating stability issues consists of using many time-steps integration in the NN construction and use, as NeuralODE performs [17]. By enforcing during the training stage the reproduction constraints in the loss function, after having performed many time steps, the stability constraint is implicitly enforced.

The main issue related to that procedure is the derivative of the loss L with respect to the network parameters, because of the fact that the loss in evaluated after integrating in time many time-steps. This calculation can be efficiently performed by integrating backward the adjoint problem for calculating the adjoint $\mathbf{a} = \partial L/\partial \mathbf{x}$, that as derived in a very intuitive way in [18] follows

$$\frac{d\mathbf{a}}{dt} = -\mathbf{a}^T \frac{\partial f(\mathbf{x}(t), t, \theta)}{\partial \mathbf{x}},$$

from which the searched gradient can be evaluated according to:

$$\frac{dL}{d\theta} = -\int_{t_1}^{t_0} \mathbf{a}^T \frac{\partial f(\mathbf{x}(t), t, \theta)}{\partial \theta}.$$

20.12 Transformers

When data consists of sequences, as it is the case when considering time series or language (very common in natural language processing, NLP), recurrent neural networks were widely considered.

20.12 Transformers

In rNN, LSTM or GRU, the data (eventually the words after an appropriate embedding) is introduced sequentially in the so-called *encoder* that uses different hidden layers where the embedded data combines with the hidden states while transiting until all the input sequence is completed (the interested reader can refer to Fig. 1 in [19]). Then, the *decoder* operates to sequentially produce the output data sequence (that could be words sequence translation).

These architectures encounter limitations related to the memory to account for long-time correlations.

The main aim of transformers is offering more flexibility is the data association, independently of its position along the time axis (or position in the words sequence).

For that purpose, the whole data or words sequence enters into the *encoder* while transiting in its different layers. As the data composing the sequence come into the encoder, all at the same time, a *positional encoding* is needed to retain the positional relations within the sequence. Moreover, when addressing words (e.g. NLP) a word encoding is also needed.

The main ingredient of *transformers* is the so-called *self attention mechanism* that aims at finding associations or correlations between the various data composing the sequence, and for that the usual *dot* product is used.

If we note by x_i, $i = 1, ..., M$ the data involved in the sequence $\{x_1, ..., x_N\}$, the self-attention mechanism (present in the different encoder layers) computes for each data-pair the weights

$$\omega_{ij} = \texttt{softmax}(\mathbf{x}_i^T \mathbf{x}_j) = \frac{e^{\mathbf{x}_i^T \mathbf{x}_j}}{\sum_k e^{\mathbf{x}_i^T \mathbf{x}_k}},$$

that ensures the positivity of the weights as well as $\sum_j \omega_{ij} = 1$, $i = 1, ..., M$.

Now, a new weighted representation of each data is obtained from the linear combination

$$\mathbf{z}_i = \sum_{j=1}^{M} \omega_{ij} \mathbf{x}_j, \quad i = 1, ..., M.$$

Thus, $\mathbf{z}_i$ becomes similar to (it approaches) the data $\mathbf{x}_j$ having the largest attention weight.

In practice, for performing the attention mechanism each data $\mathbf{x}_i$ is transformed into its associated *query* $\mathbf{q}_i$, *key* $\mathbf{k}_{j_i}$ and *value* $\mathbf{v}_i$ counterparts, from three trainable matrices $\mathbf{W}_q$, $\mathbf{W}_k$ and $\mathbf{W}_v$, from which the transformed data $\mathbf{z}_i$ reads

$$\mathbf{z}_i = \sum_{j=1}^{M} \texttt{softmax}(\mathbf{q}_i^T \mathbf{k}_j) \mathbf{v}_i, \quad i = 1, ..., M.$$

In practice instead of using a single attention mechanism, multiple attention steps perform better, leading to the so-called *multi-head attention*, whose outputs are concatenated and then reduced from an output trainable matrix.

Then, the transformed data $\mathbf{z}_i$ is sent to the different layers of the *decoder*, while a part of the target input is masked during the training phase to avoid that future data be involved into the self-attention, as discussed and illustrated in [19].

Transformers represent a promising strategy in many domains of engineering in which memory is a major protagonist, as it is the case of time-series or computational inelasticity, where the mechanical behavior depends on the whole mechanical history.

20.13 Physics-Informed Neural Networks, PINN

Numerical simulation aims at computing an unknown field $u(\mathbf{x}, t)$ defined in space $\mathbf{x}$ and time t, whose space and time dependence is governed by a parametrized partial differential equation, PDE, $\mathcal{L}(u(\mathbf{x}, t); \boldsymbol{\mu}) = F(\mathbf{x}, t; \boldsymbol{\mu})$, being $\mathcal{L}(\cdot)$ a generic linear or nonlinear differential operator involving the space and time coordinates, $\boldsymbol{\mu}$ the set of model parameters and $F(\cdot)$ the so-called loading.

At its turn, Neural Networks (NN) construct a model relating inputs and outputs, that is, NN looks for, in general, a nonlinear regression enabling making a correspondence between the inputs and outputs used in the model construction (the so-called training dataset), that is then tested in the test dataset. All this process is performed without, a priori, any knowledge, by only using the available input-output data.

However, when some knowledge exists and is available, for instance the just referred PDE, one is tempted to approximate the values of the unknown field $u(\mathbf{x}, t)$, in different points and time instants, by using a NN, while enforcing at different collocation points the fulfillment of the PDE. This gives rise to the so-called Physics-informed NN, PINN.

In order to obtain the NN connexion weights and other parameters, in the nonparametric setting, the loss function considers the residual of the PDE in those collocation points and times, that needs the NN differentiation (by using automatic differentiation [20]), as well as the residual of the initial and boundary conditions on the essential field u. Thus, the regression could be created without the need of any data. If some information is missing (e.g. boundary conditions), the available data could enable the solution procedure. In all cases the amount of data reduces drastically, thanks to the information provided by the PDE, as described and proved in [21].

This procedure can be adapted for *discovering physics* in a certain sense. One could imagine the PDE described from the sum of different terms (each one involving different partial derivatives), each one weighted by a parameter μ_i, grouped into the parameters set $\boldsymbol{\mu}$. Now, following the just described procedure, the solution and the parameters vector $\boldsymbol{\mu}$ will be obtained, even with the amount of available data remains relatively small [21].

20.14 Thermodynamics Informed Neural Networks, TINN

Thermodynamics Informed Neural Networks, TINNs, are very close to the PINN, and they are also referred to as Structure Preserving Neural Networks, SPNN. First, we restrict our discussion to the Hamiltonian case, the dissipative one will be addressed just after.

20.14 Thermodynamics Informed Neural Networks, TINN

The Hamilton's principle of least action states that the motion of an arbitrary mechanical system occurs in such a way that the so-called *action*, is minimized. The action is an integral involving the Lagrangian, and is minimized by invoking its stationarity for any possible variation of the configuration of the system, with given an initial and final configuration of the system.

The associated Euler-Lagrange equations of a system composed of N particles, with the state space defined by their positions and velocities, q_i and $\dot{q}_i$ respectively, reads

$$\frac{\partial \mathcal{L}}{\partial q_i} - \frac{\partial}{\partial t}\left(\frac{\partial \mathcal{L}}{\partial \dot{q}_i}\right) = 0, \quad i = 1, \ldots, N. \tag{20.4}$$

The so-called *Legendre's transform* applied to the Lagrangian, allows to define the momentum p_i

$$p_i = \frac{\partial \mathcal{L}}{\partial \dot{q}_i}, \tag{20.5}$$

and the associated Hamiltonian $\mathcal{H}(q_1, \ldots, q_N, p_1, \ldots, p_N)$

$$\mathcal{H}(q_1, \ldots, q_N, p_1, \ldots, p_N) = \left(\sum_{i=1}^{N} p_i \dot{q}_i\right) - \mathcal{L}(q_1, \ldots, q_N, \dot{q}_1, \ldots, \dot{q}_N),$$

that is no more than the total energy, from which the so-called canonical equations that govern the state time evolution, can be derived:

$$\forall i \in [1, \ldots, N] \begin{cases} \dot{q}_i = \frac{\partial \mathcal{H}}{\partial p_i} \\ \dot{p}_i = -\frac{\partial \mathcal{H}}{\partial q_i} \end{cases},$$

major protagonists of the developments and discussion that follows.

For a given scalar variable $\mathcal{U}$ depending on positions and momenta, i.e. $\mathcal{U}(q_1, \ldots, q_N, p_1, \ldots, p_N)$, its time derivative reads

$$\frac{d\mathcal{U}}{dt} = \sum_{i=1}^{N}\left(\frac{\partial \mathcal{U}}{\partial q_i}\dot{q}_i + \frac{\partial \mathcal{U}}{\partial p_i}\dot{p}_i\right),$$

that taking into account the canonical equation writes,

$$\frac{d\mathcal{U}}{dt} = \sum_{i=1}^{N}\left(\frac{\partial \mathcal{U}}{\partial q_i}\frac{\partial \mathcal{H}}{\partial p_i} - \frac{\partial \mathcal{U}}{\partial p_i}\frac{\partial \mathcal{H}}{\partial q_i}\right).$$

Using the so-called Poisson brackets, this can be written as

$$\frac{d\mathcal{U}}{dt} = \{\mathcal{U}, \mathcal{H}\}.$$

Energy conservation becomes a direct consequence,

$$\frac{d\mathcal{H}}{dt} = \sum_{i=1}^{N} \left(\frac{\partial \mathcal{H}}{\partial q_i} \frac{\partial \mathcal{H}}{\partial p_i} - \frac{\partial \mathcal{H}}{\partial p_i} \frac{\partial \mathcal{H}}{\partial q_i} \right) = 0.$$

In the case of a system composed of a single particle, as for instance a linear oscillator consisting of a mass m, attached to a spring of length l and stiffness k, the canonical equations can be expressed in a simple matrix form

$$\begin{pmatrix} \dot{q} \\ \dot{p} \end{pmatrix} = \begin{pmatrix} 0 & 1 \\ -1 & 0 \end{pmatrix} \begin{pmatrix} \frac{\partial \mathcal{H}}{\partial q} \\ \frac{\partial \mathcal{H}}{\partial p} \end{pmatrix}, \tag{20.6}$$

with the Hamiltonian given by $\mathcal{H} = \frac{1}{2m} p^2 + \frac{1}{2} k (q - l)^2$.

For systems composed of many particles, we define the state vector $\mathbf{z}$ as the one containing the state variables, i.e. $\mathbf{z} = (q_1, p_1, q_2, p_2, ...)$, whose time evolution reads

$$\dot{\mathbf{z}} = \mathbf{L} \nabla_{\mathbf{z}} \mathcal{H},$$

where $\mathbf{L}$ is the skew-symmetric Hamiltonian (symplectic) matrix and $\nabla_{\mathbf{z}}$ denotes the gradient in the phase space, with $\mathbf{L} = \text{diag}(\mathcal{L})$ and with the one-particle Hamiltonian matrix $\mathcal{L}$ given by

$$\mathcal{L} = \begin{pmatrix} 0 & 1 \\ -1 & 0 \end{pmatrix}.$$

For learning Hamiltonian dynamics, different possibilities exist:

- As the Hamiltonian depends on the state variables, i.e. $\mathcal{H}(\mathbf{z})$, one can assume $\nabla_{\mathbf{z}} \mathcal{H} = \mathbf{A} \mathbf{z}$, where $\mathbf{A}$ could depend on the state, $\mathrm{i} \mathbf{A}(\mathbf{z})$. Then, from the available data (state at different times) $\mathbf{z}(t)$, $\mathbf{LA}$ can be learned, subjected to the usual constraints ($\mathbf{L}$ skew-symmetric).
- When the considered system is very large, a state reduction becomes appealing. Among the diversity of reduction techniques, auto-encoders are very valuable, for making possible the complexity reduction, prior to use regressions in the latent space [22].
- Sometimes, data can me mapped into a space (usually of higher dimension) with the constraint of recovering there a linear dynamics, enabling the use of DMD for instance as model learner. This procedure is the one considered by the Koopman operator theory.
- Another valuable route, aligned with the PINN rationale, consists of looking for the Hamiltonian expression itself. Thus, a NN can be used for constructing the regression $\mathcal{H}(\mathbf{z})$. Then, the NN is trained with the available data $\mathbf{z}(t)$, that allows computing the state time derivative $\dot{\mathbf{z}}(t)$, and then the thermodynamic consistency is enforced from the loss function $\|\dot{\mathbf{z}} - \mathbf{L} \nabla_{\mathbf{z}} \mathcal{H}\|$, that drives the NN construction, i.e., the calculation of the NN parameters.

If we face a more general dissipative case, a more general formulation will be needed. This case has been solved by resorting to *metriplectic* formulations, of which the GENERIC formalism is maybe its most popular example. In a nutshell, the *General Equation for Non-Equilibrium Reversible-Irreversible Coupling*, GENERIC, formalism establishes a general description of the time evolution of the variables describing the system under scrutiny. These variables will be denoted by $\mathbf{z}_t \equiv \mathbf{z}(t)$.

GENERIC postulates an evolution of these variables as

$$\dot{\mathbf{z}}_t = \mathbf{L}(\mathbf{z}_t)\nabla E(\mathbf{z}_t) + \mathbf{M}(\mathbf{z}_t)\nabla S(\mathbf{z}_t), \quad \mathbf{z}(0) = \mathbf{z}_0,$$

where $\mathbf{L}$ is the Poisson or symplectic matrix, responsible for the Hamiltonian (thus, reversible) part of the evolution of the system. E plays the same role of $\mathcal{H}$ in Hamiltonian systems and represents the energy of the system. $\mathbf{M}$ is the so-called the friction matrix, that takes into account the irreversible part of the evolution of the system. In turn, S is the so-called Massieu potential and represents the entropy of the system at a given scale of description given by variables $\mathbf{z}$. Different levels of description of the system give rise to different entropy potentials.

To comply with the laws of thermodynamics, the following restrictions must hold: $\mathbf{L}(\mathbf{z}) \cdot \nabla S(\mathbf{z}) = \mathbf{0}$ and $\mathbf{M}(\mathbf{z}) \cdot \nabla E(\mathbf{z}) = \mathbf{0}$, often referred to as degeneracy conditions. To fulfill these conditions, it is enough to choose $\mathbf{L}$ to be skew-symmetric (it defines a Poisson bracket) and $\mathbf{M}$ symmetric, positive semi-definite.

TINNs proceed by the identification of matrices $\mathbf{L}$ and $\mathbf{M}$ from data—for $\mathbf{L}$ this is often straightforward, many examples are known in the literature—and the particular structure of the gradients of energy and entropy (Hamiltonian and dissipative parts of the constitutive equations, respectively). This identification is done from (pseudo-)experimental data and shows promising results [23–25].

20.15 Generative Adversarial Network, GAN

Generative Adversarial Networks, GANs, sketched in Fig. 20.7, learn by trying to differentiate real data from noisy data. It contains two NN, the first called the *Generator* and the second the *Discriminator*. The *generator* tries to mislead the *Discriminator*, and the last to unveil the data created by the generator.

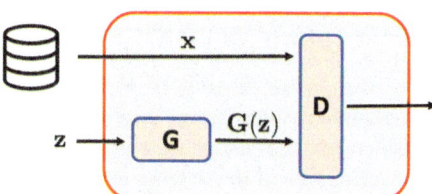

Fig. 20.7 GAN global architecture

The *Generator* generates random inputs from a latent space, while computing its parameters θ^G from the gradient of its cost function J^G; whereas the *Discriminator* uses as input the real data as well as the one generated by the *Generator* to optimize its parameters θ^D to discriminate true and false data, by diminishing its associated cost function J^D by using a gradient minimizer. Both processes operate simultaneously.

With the true and generated data are noted respectively by $\mathbf{x}$ and $\mathbf{G}(\mathbf{z})$, labelled respectively with 1 and 0, then the *Discriminator* tries to produce $D(\mathbf{x}) = 1$ and $D(\mathbf{G}(\mathbf{z})) = 0$ respectively. The loss could be defined as

$$J^D = \mathbb{E}_x\{\log D(\mathbf{x})\} + \mathbb{E}_z\{\log(1 - D(\mathbf{G}(\mathbf{z})))\},$$

where $\mathbb{E}$ refers to the expectations of the true and generated data distributions.

Concerning the *Generator* its cost function consists of $J^G = -J^D$, and defines the so-called minimax game. An improved version considers

$$J^G = -\mathbb{E}_z\{\log D(\mathbf{G}(\mathbf{z}))\}.$$

Thus, when considering a sampling consisting of m true data $\mathbf{x}_i$, $i = 1, ..., m$; and m generated data $\mathbf{G}(\mathbf{z}_i)$, $i = 1, ..., m$; the gradients considered in the training read

$$\begin{cases} \nabla_{\theta^D} \frac{1}{m} \sum_{i=1}^{m} \{\log D(\mathbf{x}_i) + \log(1 - D(\mathbf{G}(\mathbf{z}_i)))\} \\ \nabla_{\theta^G} \frac{1}{m} \sum_{i=1}^{m} \log D(\mathbf{G}(\mathbf{z}_i)) \end{cases}.$$

The same rationale can be applied when considering time series, leading in this case to the so-called TimeGAN. TimeGAN consists of four main blocks: the standard *Generator* and *Discriminator* as well as two specific components, an *Embedding* and a *Recovering* functions. The main interest of the autoencoding (embedding) is extracting the latent space, such that the adversarial network operates in the reduced (latent) space. The latent dynamics of both real and synthetic data are synchronized through a supervised loss. Additional details can be found in [26].

GANs were also extended to conditional models, where both, the *Generator* and the *Discriminator*, are conditioned on some extra-information (e.g. labels) [27].

20.16 Nonlinear Autoregressive Network with Exogenous Inputs, NARX

The nonlinear autoregressive exogenous model, NARX, [28], applied to time series, considers that the current state $y(t)$ depends on the past values as well as on the current and past values of the driving (exogenous) time series $u(t)$. Thus, with $y_n \equiv y(t =$

$n\Delta t$) the variable of interest and $u_n \equiv u(t = n\Delta t)$ the external driving variable, then the model reads

$$y_n = y_n(y_{n-1}, y_{n-2}, ..., u_n, u_{n-1}, u_{n-2}, ...) + \epsilon_n,$$

with ϵ_n the so-called noise.

The nonlinear function can be determined by suing polynomials or a Neural Network –NN– based regression. The great success of using NN is again, the universal approximation theorem.

When using NN, the so-called Tapped-Delay Line, TDL, extracts the m past values of the external series $u(t)$: $u_n, ..., u_{n-m+1}$. A second TDL contains the m previous values of the time series $y(t)$. Both TDLs are used as inputs of a NN, being y_n its output. y_n is then re-injected into the TDL of the y-values. To assimilate the just computed value into the $y-$TPL, the last value (the oldest) is removed while all the previous values move back one position.

20.17 Multimodal Learning Based on Boltzmann Machines, BM

In many applications, data come from different sources (images, texts, sound, ...), and they should be convenable combined. The different data typologies having different statistical properties, the relations between them must be extracted. Multimodal learning involves, other than the stochastic visible variables, the so-called stochastic hidden variables, to be extracted from the visible ones.

The usual architecture of such a learners involves a Boltzmann machine –BM– (os its deep counterpart –DBM–) associated to each mode and another that joins them.

BM [29, 30] involve interconnected binary stochastic visible and hidden units, allowing the definition of the so-called energy function (like in statistical mechanics) from which the distribution over the visible and hidden units is derived.

In what follows we focus in its simplest typology, the co-called Restricted Boltzmann Machine –RBM–, that consist of a undirected graphical model (with each visible unit connected with all the hidden ones), with D visible scholastic variables and F hidden, organized respectively in the vectors $\mathbf{v} \in \{0, 1\}^D$ and $\mathbf{h} \in \{0, 1\}^F$. Figure 20.8 depicts a RBM composed of four visible and four hidden units. The unsupervised training follows from the procedure described below.

The RBM energy $E : \{0, 1\}^D \times \{0, 1\}^F \to \mathbb{R}$, reads

$$E(\mathbf{v}, \mathbf{h}; \theta) = -\sum_{i=1}^{D}\sum_{j=1}^{F} W_{ij} v_i h_j - \sum_{i=1}^{D} b_i v_i - \sum_{j=1}^{F} a_j h_j,$$

where θ refers to the model parameters $\theta = \{W_{ij}, b_i, a_j\}$, $i = 1, ..., D$ and $j = 1, ..., F$.

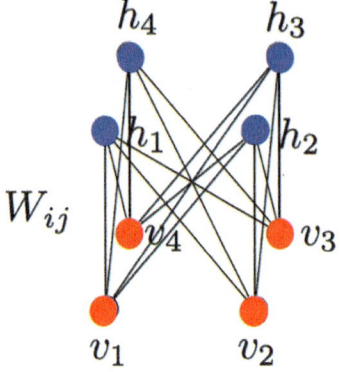

Fig. 20.8 Restricted Boltzmann Machine with four visible (red) and four hidden (blue) units

The energy definition allows defining the joint distribution $P(\mathbf{v}, \mathbf{h}; \boldsymbol{\theta})$

$$P(\mathbf{v}, \mathbf{h}; \boldsymbol{\theta}) = \frac{\exp\{-E(\mathbf{v}, \mathbf{h}; \boldsymbol{\theta})\}}{\mathcal{Z}(\boldsymbol{\theta})},$$

with $\mathcal{Z}(\boldsymbol{\theta})$ the normalizing coefficient,

$$\mathcal{Z}(\boldsymbol{\theta}) = \sum_{\mathbf{v}} \sum_{\mathbf{h}} \exp\{-E(\mathbf{v}, \mathbf{h}; \boldsymbol{\theta})\}.$$

The conditional distributions are factorized as follows:

$$\begin{cases} P(\mathbf{h}|\mathbf{v}; \boldsymbol{\theta}) = \prod_{j=1}^{F} p(h_j|\mathbf{v}; \boldsymbol{\theta}) \\ P(\mathbf{v}|\mathbf{h}; \boldsymbol{\theta}) = \prod_{i=1}^{D} p(v_i|\mathbf{h}; \boldsymbol{\theta}) \end{cases},$$

where the right members are approximated on the RBM network according to

$$\begin{cases} p(h_j|\mathbf{v}; \boldsymbol{\theta}) = \sigma\left(\sum_{i=1}^{D} W_{ij} v_i + a_j\right) \\ p(v_i|\mathbf{h}; \boldsymbol{\theta}) = \sigma\left(\sum_{j=1}^{F} W_{ij} h_j + b_i\right) \end{cases},$$

with $\sigma(\cdot)$ the usual activation function (e.g. the logistic function).

If we assume N observations of the visible variables (visible units) $\{\mathbf{v}_1, ..., \mathbf{v}_N\}$, the distribution reads

$$P(\mathbf{v}_1, ..., \mathbf{v}_N; \boldsymbol{\theta}) = \prod_{n=1}^{N} \left\{ \frac{\exp\{-\mathbb{E}_h(E(\mathbf{v}_n, \mathbf{h}; \boldsymbol{\theta}))\}}{\mathcal{Z}(\boldsymbol{\theta})} \right\},$$

20.17 Multimodal Learning Based on Boltzmann Machines, BM

or

$$P(\mathbf{v}_1, ..., \mathbf{v}_N; \boldsymbol{\theta}) = \frac{\exp\left\{-\sum_{n=1}^{N} \mathbb{E}_h(E(\mathbf{v}_n, \mathbf{h}; \boldsymbol{\theta}))\right\}}{\mathcal{Z}^N(\boldsymbol{\theta})}. \tag{20.7}$$

The logarithm of Eq. (20.7) reads

$$\ln(P) = -\sum_{n=1}^{N} \mathbb{E}_h(E(\mathbf{v}_n, \mathbf{h}; \boldsymbol{\theta})) - N \ln \mathcal{Z}(\boldsymbol{\theta}) = \mathcal{T}_d + \mathcal{T}_m, \tag{20.8}$$

where $\mathcal{T}_d$ and $\mathcal{T}_m$ refers the a data and model contributions respectively:

$$\begin{cases} \mathcal{T}_d = -\sum_{n=1}^{N} \mathbb{E}_h(E(\mathbf{v}_n, \mathbf{h}; \boldsymbol{\theta})) \\ \mathcal{T}_m = -N \ln \mathcal{Z}(\boldsymbol{\theta}) \end{cases}.$$

The maximization of the $\ln(P)$ given by Eq. (20.8) implies the vanishment of its derivative with respect to the approximation weights W_{ij}. The derivatives of both contributing terms are given below. The first results

$$\frac{\partial \mathcal{T}_d}{\partial W_{ij}} = -\sum_{n=1}^{N} \mathbb{E}_h\left(\frac{\partial E(\mathbf{v}_n, \mathbf{h}; \boldsymbol{\theta})}{\partial W_{ij}}\right) = \sum_{n=1}^{N} \mathbb{E}_h((v_n)_i h_j) = N \cdot \mathbb{E}(v_i h_j),$$

where in order to emphasize that the average concerns the data, $\mathbb{E}$ will be referred in the sequel as $\mathbb{E}_d$.

The derivation of the second one follows from

$$\frac{\partial \ln \mathcal{Z}}{\partial W_{ij}} = \frac{\frac{\partial \mathcal{Z}}{\partial W_{ij}}}{\mathcal{Z}} = \frac{\sum_\mathbf{v} \sum_\mathbf{h} v_i h_j \exp\{-E(\mathbf{v}, \mathbf{h}; \boldsymbol{\theta})\}}{\mathcal{Z}}(\boldsymbol{\theta}) = \mathbb{E}_m(v_i h_j),$$

with $\mathbb{E}_m$ referring the average with respect to the model. Thus, we obtain

$$\frac{\partial \mathcal{T}_m}{\partial W_{ij}} = -N \cdot \mathbb{E}_m(v_i h_j).$$

Thus, we obtain finally

$$\frac{1}{N} \frac{\partial \ln P}{\partial W_{ij}} = \mathbb{E}_d(v_i h_j) - \mathbb{E}_m(v_i h_j),$$

whose vanishment drives the training process.

The rationale can be extended for a variety of configurations (connections), choice of the energies, and number of hidden layers. When addressing multi-modal learning, the Boltzmann machines for each mode join in another that collect and join them. A schema for two modes is given in Fig. 20.9.

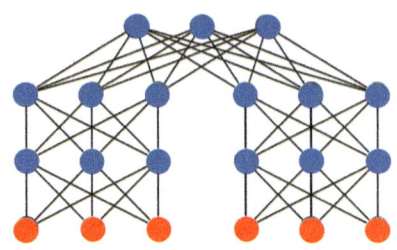

Fig. 20.9 Scheme of a two modes multi-modal learner composes of three deep Boltzmann machines

References

1. I. Goodfellow, Y. Bengio, A. Courville, *Deep Learning* (MIT Press, Cambridge, 2016)
2. M. Nielsen, Neural networks and deep learning (2019). http://neuralnetworksanddeeplearning.com/chap4.html
3. T. Chen, H. Chen, Approximations of continuous functionals by neural networks with application to dynamic systems. IEEE Trans. Neural Netw. **4**(6), 910–918 (1993)
4. T. Chen, H. Chen, Universal approximation to nonlinear operators by neural networks with arbitrary activation functions and its application to dynamical systems. IEEE Trans. Neural Netw. **6**(4), 911–917 (1995)
5. L. Lu, P. Jin, G.E. Karniadakis, DeepONet: learning nonlinear operators for identifying differential equations based on the universal approximation theorem of operators (2020). arXiv:1910.03193v3
6. R. Venkatesan, B. Li, *Convolutional Neural Networks in Visual Computing: A Concise Guide* (CRC Press, 2017)
7. M. Bronstein, J. Bruna, T. Cohen, P. Velickovic, Geometric Deep Learning, Grids, Groups, Graphs, Geodesics and Gauges. https://arxiv.org/abs/2104.13478
8. Q. Hernandez, A. Badias, F. Chinesta, E. Cueto. Themodynamic-informed graph neural networks. IEEE Trans. Artif. Intell.: Spec. Issue Phys.-Inf. Mach. Learn. https://doi.org/10.1109/TAI.2022.3179681
9. T.W. Hughes, I.A.D. Williamson, M. Minkov, S. Fan, Wave physics as an analog recurrent neural network. Sci. Adv. **5**(12), easy6946 (2019)
10. G. Zhou, J. Wu, C. Zhang, Z. Zhou, Minimal gated unit for recurrent neural networks. Int. J. Autom. Comput. **13**(3), 226–234 (2016)
11. B. Lahouari, Development of LSTM networks for predicting viscoplasticity with effects of deformation. Strain Rat. Temp. Hist. J. Appl. Mech. **88**(7), 071008 (2021)
12. H. Luo, M. Huang, Z. Zhou, Integration of multi-Gaussian fitting and LSTM neural networks for health monitoring of an automotive suspension component. J. Sound Vib. **428**, 87–103 (2018)
13. S. Hochreiter, J. Schmidhuber, Long short-term memory. Neural Comput. **9**(8), 1735–1780 (1997)
14. C. Ghnatios, X. Kestelyn, G. Denis, V. Champaney, F. Chinesta, Learning data-driven stable corrections of dynamical systems-application to the simulation of the top-oil temperature evolution of a power transformer. Energies **16**, 5790 (2023)
15. J. Chung, C. Gulcehre, K.H. Cho, Y. Bengio, Empirical evaluation of gated recurrent neural networks on sequence modeling. arXiv:1412.3555
16. F. Xue, Q. Li, X. Li, The combination of circle topology and leaky integrator neurons remarkably improves the performance of echo state network on time series prediction. PLoS One 12(7), e0181816 (2017)
17. R.T.Q. Chen, Y. Rubanova, J. Bettencourt, D.K. Duvenaud, Neural ordinary differential equations. Adv. Neural Inf. Process. Syst. **31** (NeurIPS 2018)

References

18. I. Schurov, Adjoint State Method, Backpropagation and Neural ODEs. https://ilya.schurov.com/post/adjoint-method/
19. S. Ahmed, I.E. Nielsen, A. Tripathi, S. Siddiqui, G. Rasool, R.P. Ramachandran, Transformers in time-series analysis: a tutorial. Circuits Syst. Signal Process (2023). https://doi.org/10.1007/s00034-023-02454-8
20. A.G. Baydin, B.A. Pearlmutter, A.A. Radul, J.M. Siskind, Automatic differentiation in machine learning: a survey. J. Mach. Learn. Res. **18**, 1–43 (2018)
21. M. Raissi, P. Perdikaris, G.E. Karniadakis, Physics-informed neural networks: a deep learning framework for solving forward and inverse problems involving nonlinear partial differential equations. J. Comput. Phys. **378**, 686–707 (2019)
22. B. Lusch, J.N. Kutz, S.L. Brunton, Deep learning for universal linear embeddings of nonlinear dynamics. Nat. Commun. **9**, 4950 (2018)
23. D. Gonzalez, J.V. Aguado, E. Cueto, E. Abisset-Chavanne, F. Chinesta, kPCA-based parametric solutions within the PGD framework. Arch. Comput. Methods Eng. **25**, 69–86 (2018)
24. Q. Hernadez, D. Gonzalez, F. Chinesta, E. Cueto, Learning non-Markovian physics from data. J. Comput. Phys. **428**, 109982 (2021)
25. Q. Hernandeza, A. Badias, D. Gonzalez, F. Chinesta, E. Cueto, Deep learning of thermodynamics-aware reduced-order models from data. J. Comput. Phys. **426**, 109950 (2021)
26. J. Yoon, D. Jarrett, M. van der Schaar, Time-Series Generative Adversarial Networks Conference: Neural Information Processing Systems (NeurIPS) (2019). https://papers.nips.cc/paper/8789-time-series-generative-adversarial-networks.pdf
27. M. Mirza, S. Osindero, Conditional Generative Adversarial Nets (2014). https://arxiv.org/abs/1411.1784v1
28. S.A. Billings, Nonlinear System Identification: NARMAX Methods in the Time, Frequency, and Spatio-Temporal Domains (Wiley, 2013)
29. R. Salakhutdinov, G. Hinton, Deep Boltzmann machines, in *Appearing in Proceedings of the 12th International Conference on Artificial Intelligence and Statistics (AISTATS)*, Clearwater Beach, Florida, USA. Volume 5 of JMLR (2009)
30. N. Srivastava, R. Salakhutdinov, Multimodal learning with deep Boltzmann machines. J. Mach. Learn. Res. **15**, 2949–2980 (2014)

Open Access This chapter is licensed under the terms of the Creative Commons Attribution-NonCommercial-NoDerivatives 4.0 International License (http://creativecommons.org/licenses/by-nc-nd/4.0/), which permits any noncommercial use, sharing, distribution and reproduction in any medium or format, as long as you give appropriate credit to the original author(s) and the source, provide a link to the Creative Commons license and indicate if you modified the licensed material. You do not have permission under this license to share adapted material derived from this chapter or parts of it.

The images or other third party material in this chapter are included in the chapter's Creative Commons license, unless indicated otherwise in a credit line to the material. If material is not included in the chapter's Creative Commons license and your intended use is not permitted by statutory regulation or exceeds the permitted use, you will need to obtain permission directly from the copyright holder.

Chapter 21
Other Machine Learning Techniques

21.1 Code2Vect

Very often, data comes in different formats but still needs to be integrated in some form of decision making. *Code2Vect* [1] is a technique that maps heterogeneous data (discrete, categorial, ...) to a vector space. This allows us to define a scalar product and therefore a distance. This is a critical ingredient to quantify distance as sketched in Fig. 21.1, distance that many clustering, classification and regression techniques employ.

Points in the original space (called the space of representation) consist of M labelled arrays (in general, they cannot being considered vectors) composed of P entries which we denote by $\mathbf{x}_i$. Their images in the vector space are $\mathbf{z}_i \in \mathbb{R}^p$, with $p \ll P$. These are actual vectors, subjected to the rules of coordinate transformation so that the vector space is equipped with a scalar product and a norm, that lead in general to an Euclidean distance. This mapping is given by the $p \times P$ matrix $\mathbf{W}$, that satisfies $\mathbf{z} = \mathbf{W}\mathbf{x}$. Here, both the components of $\mathbf{W}$ and the images $\mathbf{z}_i \in \mathbb{R}^p$, $i = 1, \ldots, M$, must be obtained. Each point $\mathbf{z}_i$ is assigned the same label of its origin point $\mathbf{x}_i$, denoted by y_i.

The objective is to place points $\mathbf{z}_i$ at an Euclidian distance with respect other point $\mathbf{z}_j$ so that it scales with their output difference, i.e.

$$(\mathbf{W}(\mathbf{x}_i - \mathbf{x}_j)) \cdot (\mathbf{W}(\mathbf{x}_i - \mathbf{x}_j)) = \|\mathbf{z}_i - \mathbf{z}_j\|^2 = |y_i - y_j|,$$

where the coordinates of one of the points can be arbitrarily chosen. This gives $(M^2 - M)/2$ relations to find out the $p \times P + M \times p$ unknowns.

Linear mappings constitute an obvious choice, but they posses obvious limitations and do not work well in nonlinear settings. The nonlinear mapping $\mathbf{W}(\mathbf{x})$, with a general polynomial form

$$\mathbf{W}(\mathbf{x}) = \sum_{k=1}^{K} \mathbb{W}_k \mathcal{P}_k(\mathbf{x}),$$

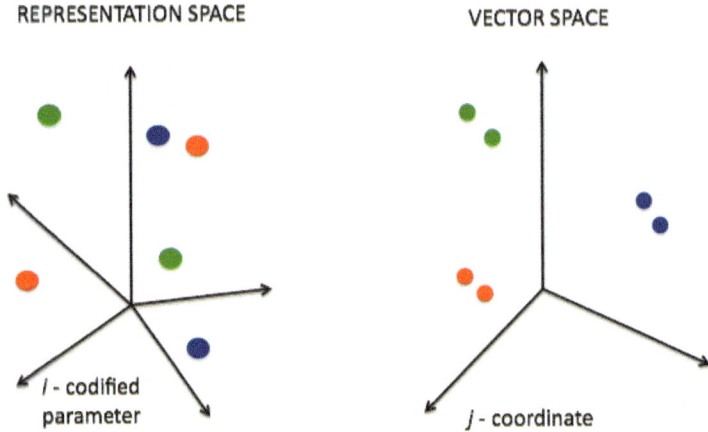

Fig. 21.1 *Code2Vect* input space (left) and target vector space (right)

where $\mathbb{W}_k$ are $p \times P$ matrices and $\mathcal{P}(\mathbf{x})$ a polynomial basis, constitutes an appealing choice. This mapping could be expressed, for instance, in a separate form, inspired by the PGD.

21.2 Sparse Identification for Nonlinear Dynamical Systems, SINDy

Sparse identification is widely considered to learn nonlinear governing laws from data, [2]. It assumes the existence of a dynamical system governed by a law of the form

$$\frac{d\mathbf{x}}{dt} = \mathbf{f}(\mathbf{x}, t),$$

where the vector $\mathbf{x}(t) \in \mathbb{R}^D$ denotes the state of the system at time t.

In the previous expression, $\mathbf{f}$ that governs the evolution of the system is assumed unknown. The main ingredient of the method is the assumption of a parsimonious model, that is, that the sought $\mathbf{f}$ is composed by few terms depending on $\mathbf{x}$.

If the snapshots matrix $\mathbf{X}$ is defined as

$$\mathbf{X} = \begin{pmatrix} \mathbf{x}^T(t_1) \\ \mathbf{x}^T(t_2) \\ \vdots \end{pmatrix}$$

SINDy constructs a library of linear and nonlinear functions involving $\mathbf{X}$

$$\Theta(\mathbf{X}) = \begin{bmatrix} | & | & | & & | & & | & \\ 1 & \mathbf{X} & \mathbf{X}^2 & \cdots & \sin(\mathbf{X}) & \cdots & \exp(\mathbf{X}) & \cdots \\ | & | & | & & | & & | & \end{bmatrix}.$$

According to the parsimony assumption, the final step consist in a sparse regression so as to obtain the terms in $\Xi(\mathbf{X})$ that are present in the governing of the dynamics of the system: $\dot{\mathbf{X}} = \Theta(\mathbf{X})\Xi$, where Ξ contains the sparse reduced vectors. Sparsity is enforced by solving the identification problems by using the Lasso algorithm previously introduced.

21.3 Sparse PGD Based Regressions, sPGD

The main drawbacks of usual linear regressions making use of polynomial approximation bases were pointed out in the previous section. Here we review an alternative method that allows us to obtain sparse approximations in high-dimensional spaces. Without loss of generality, let us assume that the unknown objective function $f(x, y)$ lives in $\mathbb{R}^2$—this will allow us to represent graphically some concepts—and that is to be recovered from sparse data.

Consider the classic Galerkin projection

$$\int_\Omega w(x, y)\, (u(x, y) - f(x, y))\, dxdy = 0,$$

where $\Omega \subset \mathbb{R}^2$. Following the rationale behind Proper Generalized Decompositions, we express the function $u(x, y)$ in a separated form, and at iteration n with the approximation $u^{n-1}(x, t)$ known, looks for an improved approximation

$$u^n(x, y) = u^{n-1}(x, y) + X_n(x)Y_n(y),$$

with

$$u^{n-1}(x, y) = \sum_{k=1}^{n-1} X_k(x)Y_k(y).$$

Note that the product of the test function $w(x, y)$ by the sought function $f(x, y)$ can only be evaluated at a few locations (these corresponding to the sampling points). Since information is just known at M of these points (x_i, y_i), $i = 1, \ldots, M$, we use a collocation scheme [3]

$$w(x, y) = \left(X^*(x)Y_n(y) + X_n(x)Y^*(y)\right) \sum_{i=1}^{M} \delta(x_i, y_i),$$

giving rise to

$$\int_\Omega \left(X^*(x)Y_n(y) + X_n(x)Y^*(y)\right) \left(\sum_{i=1}^M \delta(x_i, y_i)\right) (u(x,y) - f(x,y))\, dxdy = 0.$$

The most appealing property of the just proposed procedure is that because of the fact of using an alternated direction rank-one updating in each direction, the fact of having M available data enables richer approximations with respect to usual linear regressions. Obviously, sparsity can be enforced using an adequate norm (L1 or the Lasso regularization) as later discussed in Chap. 27.

Note that we have not yet specified anything about the basis in which to express each of the one-dimensional modes. In general, globally-supported one-dimensional interpolations should be preferred to avoid rank deficiency. This may occur when the support of a local approximation function does not contain any of the sampled data.

Kriging approximation constitutes an appealing choice, since it ensures accuracy and avoids spurious oscillations (overfitting). Standard polynomial bases can also be considered, provided that we establish a procedure to adapt the approximation degree, as discussed later in Chap. 27.

21.4 The Koopman Operator

The so-called Koopman operator is a linear operator aiming at addressing nonlinear dynamical systems. If we consider the nonlinear dynamical system, with the state $\mathbf{x} \in \mathbb{R}^N$ expressed at a certain time $t_k = k\Delta t$ as $\mathbf{x}_k$, the nonlinear dynamical system discrete form reads:

$$\mathbf{x}_{k+1} = \mathbf{f}(\mathbf{x}_k).$$

We define the observation $g : \mathbf{x} \to g(\mathbf{x}) \in \mathbb{R}$ (vector valued observables will be considered later). The Koopman operator $\mathcal{K}$ is defined from:

$$\mathcal{K}g(\mathbf{x}) \equiv g(\mathbf{f}(\mathbf{x})).$$

The linear character of the Koopman operator allows its eigendecomposition,

$$\mathcal{K}\varphi_j(\mathbf{x}) = \lambda_j \varphi_j(\mathbf{x}), \quad j = 1, ..., \infty.$$

If we consider now a vector valued observable $\mathbf{g} : \mathbb{R}^N \to \mathbb{R}^M$. Each component of $\mathbf{g}$ can be expanded in the basis φ_j, according to

$$\mathbf{g}(\mathbf{x}) = \sum_{j=1}^\infty \varphi_j \mathbf{v}_j,$$

where the Koopman modes $\mathbf{v}_j$ come from the internal product: $(\mathbf{v}_j)_i = \langle \varphi_j(\mathbf{x}), \mathbf{g}_i \rangle$.

Thus, from the Koopman definition we obtain the following representation

$$\mathcal{K}\mathbf{g}(\mathbf{x}) = \mathbf{g}(\mathbf{f}(\mathbf{x})) = \sum_{i=1}^{\infty} \lambda_j \varphi_j(\mathbf{x}) \mathbf{v}_j,$$

where the Koopman eigenvalues provide the growth of each Koopman mode $\mathbf{v}_j$.

A significant amount of recent works concern the application of the adequate observables g in order to uncover a Koopman operator that describes the nonlinear vector field.

The dynamic mode decomposition, DMD, described in the next section, uses

$$\begin{cases} \mathbf{y}_k = \mathbf{g}(\mathbf{x}_k) \\ \mathbf{z}_k = \mathbf{g}(\mathbf{f}(\mathbf{x}_k)) \end{cases},$$

that allows calculating the linear model $\mathbf{A}$ from $\mathbf{A} = \mathbf{Z} \cdot \mathtt{pinv}(\mathbf{Y})$, where $\mathbf{Y} = [\mathbf{y}_1 \ \mathbf{y}_2 \ \cdots]$, $\mathbf{Z} = [\mathbf{z}_1 \ \mathbf{z}_2 \ \cdots]$, and $\mathtt{pinv}(\cdot)$ refers to the pseudo-inverse.

In [4], the authors took advantage of this subtle connexion between the Koopman operator and the DMD, and used autoencoders for facilitating the application of DMD in the latent space (the same rationale was applied by the same authors in the context of sparse identification). In that case the training of the autoencoder-based embedding, and the DMD model identification, are performed simultaneously, enforcing the embedding to learn the Koopman space where the dynamics becomes linear.

21.5 Dynamic Mode Decomposition, DMD

Dynamic Mode Decomposition (DMD) [5, 6] aims at identifying the dynamics $\dot{\mathbf{x}} = \mathbf{f}(\mathbf{x})$, where $\mathbf{x}(t) \in \mathbb{R}^D$ represents the vector of the system state variables (usually large) and $\mathbf{f}$ represents the sought dynamics.

The discrete *flow map* of the system reads

$$\mathbf{x}_{n+1} = \mathbf{f}(\mathbf{x}_n),$$

with $\mathbf{x}_n = \mathbf{x}(t = n\Delta t)$.

DMD looks for a locally linear approximation to the dynamics, i.e. $\mathbf{x}_{n+1} = \mathbf{A}\mathbf{x}_n$, where the model $\mathbf{A}$ comes from a least-squares minimization

$$\mathbf{A} = \arg\min_{\mathbf{A}^*} \|\mathbf{x}_{n+1} - \mathbf{A}^*\mathbf{x}_n\|_2, \ \forall n = 1, \ldots, M-1.$$

Tthere is no guaranteed optimality for $\mathbf{A}$ outside the interval defined by the M measurements, but in general, the predictions can be successfully extrapolated in the near future.

In matrix form, the method performs by constructing a matrix of snapshots in the form

$$\mathbf{X} = \begin{bmatrix} | & | & & | \\ \mathbf{x}_1 & \mathbf{x}_2 & \cdots & \mathbf{x}_{M-1} \\ | & | & & | \end{bmatrix},$$

as well as a second one very close

$$\mathbf{X}' = \begin{bmatrix} | & | & & | \\ \mathbf{x}_2 & \mathbf{x}_3 & \cdots & \mathbf{x}_M \\ | & | & & | \end{bmatrix},$$

such that

$$\mathbf{X}' \approx \mathbf{A}\mathbf{X}.$$

The sought matrix $\mathbf{A}$ will then be given by the Moore-Penrose pseudo-inverse

$$\mathbf{A} = \mathbf{X}'\mathbf{X}^{-1},$$

which is known to minimize the Frobenius norm of the residual, i.e.

$$\mathbf{A} = \arg\min_{\mathbf{A}^*} \|\mathbf{X}' - \mathbf{A}^*\mathbf{X}\|_F.$$

When the size of matrices $\mathbf{X}$ and $\mathbf{X}'$ is huge, DMD performs a POD projection of the data, trying to minimize the number of degrees of freedom with whom we parameterize the system.

To ensure stability when using the learnt dynamical system to perform time integrations, some constraints must be added when learning matrix $\mathbf{A}$ concerning its eigenvalues.

21.6 Incremental DMD, iDMD

Consider an already discretized problem $\mathbf{KU} = \mathbf{F}$, which is assumed to be linear, where $\mathbf{F}$ and $\mathbf{U}$ constitute the input and output vectors of the system, respectively. These vectors store the values of their respective fields (often of different nature) at the sampling points. We assume their dimensions to be P × 1 for both.

Very much like in the reduced-order modeling framework, it is assumed that inputs and outputs (to a certain degree of approximation) live in a subspace of dimension p, much smaller than P. It seems thus reasonable to expect the rank of $\mathbf{K}$ to be also p, even if it was thought to operate in a larger space of dimension P.

The objective is therefore to unveil a reduced order model. We summarize two of the different possibilities [7] (i) the Progressive Greedy Constructor; and (ii) the rank-p constructor.

21.6.1 Progressive Greedy Constructor, PGC

We start by considering only one datum, the pair $(\mathbf{F}_1, \mathbf{U}_1)$. The first model will obviously be of rank one, $\mathbf{K}_1 = \mathbf{F}_1 \mathbf{F}_1^T / \mathbf{F}_1^T \mathbf{U}_1$, satisfying $\mathbf{K}_1 \mathbf{U}_1 = \mathbf{F}_1$.

Assuming now the recording of a second datum, $(\mathbf{F}_2, \mathbf{U}_2)$, we can also compute its associated rank-one approximation. We proceed similarly for any new datum $(\mathbf{F}_i, \mathbf{U}_i)$: $\mathbf{K}_i = \mathbf{F}_i \mathbf{F}_i^T / \mathbf{F}_i^T \mathbf{U}_i$.

For any other $\mathbf{U}$, the model could be interpolated from the just defined rank-one models, $\mathbf{K}_i, i = 1, \ldots, M$, according to

$$\mathbf{K}|_{\mathbf{U}} \approx \sum_{i=1}^{M} \mathbf{K}_i \mathcal{P}_i(\mathbf{U}),$$

with $\mathcal{P}_i(\mathbf{U})$ the chosen interpolation functions defined in the space of the data $\mathbf{U}$.

All the previous analysis was based on the approximation of the reduced model $\mathbf{K}$. This requires the inversion of $\mathbf{K}$, that is not defined, to compute the solution $\mathbf{U} = \mathbf{K}^{-1}\mathbf{F}$. This limitation can easily be circumvented by looking for an approximation of the inverse $\mathbf{K}^{-1}$, the so-called transfer function, noted by $\mathbf{T}$, ensuring $\mathbf{U} = \mathbf{T}\mathbf{F}$, with $\mathbf{T}_i = \mathbf{U}_i \mathbf{U}_i^T / \mathbf{U}_i^T \mathbf{F}_i$.

21.6.2 Rank-p Constructor, RPC

Consider now a set of M input-output couples $(\mathbf{F}_i, \mathbf{U}_i)$, and the low-rank form $\mathbf{K}^{\mathrm{LR}}$ that can be obtained from them,

$$\mathbf{K} \approx \mathbf{K}^{\mathrm{LR}} = \sum_{j=1}^{p} \mathbf{C}_j \otimes \mathbf{R}_j = \sum_{j=1}^{p} \mathbf{C}_j \mathbf{R}_j^{\mathrm{T}}$$

where $\otimes$ denotes the usual tensor product, and $\mathbf{C}_j$ and $\mathbf{R}_j$ are *column* and *row* vectors, respectively. This form resembles the separate representation used in the PGD, the SVD (Singular Value Decomposition) or the CUR decomposition. In addition, symmetry can be enforced by ssuming $\mathbf{C}_j \mathbf{R}_j^{\mathrm{T}} + \mathbf{R}_j \mathbf{C}_j^{\mathrm{T}}$.

We now define the functional $\mathcal{E}(\mathbf{K}^{\mathrm{LR}})$ as

$$\mathcal{E}(\mathbf{K}^{\mathrm{LR}}) = \sum_{i=1}^{M} \left\| \mathbf{F}_i - \mathbf{K}^{\mathrm{LR}} \mathbf{U}_i \right\|^2.$$

If we group columns vectors $\mathbf{F}_i$ and $\mathbf{U}_i$ within matrices $\mathbb{F}$ and $\mathbb{U}$ respectively, the previous expression can be rewritten as $\mathcal{E}(\mathbf{K}^{\mathrm{LR}}) = \left\| \mathbb{F} - \mathbf{K}^{\mathrm{LR}} \mathbb{U} \right\|^2$, whose minimization results in $\mathbf{K}^{\mathrm{LR}}(\mathbb{U}\mathbb{U}^{\mathrm{T}}) = \mathbb{F}\mathbb{U}^{\mathrm{T}}$, or, equivalently, $\mathbf{K}^{\mathrm{LR}} = (\mathbb{F}\mathbb{U}^{\mathrm{T}})(\mathbb{U}\mathbb{U}^{\mathrm{T}})^{-1}$. This proves that $\mathbf{K}^{\mathrm{LR}}$ and, more particularly, its column and row vectors, correspond to the (rank-n truncated) SVD decomposition of $(\mathbb{F}\mathbb{U}^{\mathrm{T}})(\mathbb{U}\mathbb{U}^{\mathrm{T}})^{-1}$.

21.6.3 Reduced Formulation

A valuable route consists of extracting a reduced basis from vectors $\mathbf{U}_i$, by using a $P \times p$ projection matrix $\mathbf{B}$, so that vectors $\mathbf{U}_i$ can be expressed as $\mathbf{U}_i = \mathbf{B}\mathbf{u}_i$. The system will now read $\left(\mathbf{B}^T\mathbf{K}\mathbf{B}\right)\mathbf{u} = \mathbf{B}^T\mathbf{F}$, or in a more compact form, $\mathbf{k}\mathbf{u} = \mathbf{f}$. Thus, starting from data, $\mathbf{U}_i$ and $\mathbf{F}_i$, we first compute their reduced counterparts, $\mathbf{u}_i = \mathbf{B}^T\mathbf{U}_i$ and $\mathbf{f}_i = \mathbf{B}^T\mathbf{F}_i$, and then the resulting reduced model $\mathbf{k}$ from $\mathbf{k} = (\mathfrak{F}\mathfrak{U}^T)(\mathfrak{U}\mathfrak{U}^T)^{-1}$, where $\mathfrak{U} = \{\mathbf{u}_1 \cdots \mathbf{u}_M\}$ and $\mathfrak{F} = \{\mathbf{f}_1 \cdots \mathbf{f}_M\}$.

21.6.4 Noise Filtering

To avoid the overfitting effects that noisy data can induce, many possibilities exist. Our numerical solutions consider a quite simple and effective procedure. Data vectors $(\mathbf{F}_i, \mathbf{U}_i)$ are obtained with a resolution P that includes some noise.

A simple filter can be obtained by simply constructing the model, through its row and column vectors, at a coarser resolution $\tilde{P} < P$. While inputs and outputs could exhibit fast fluctuations, in a reduced-order modeling framework, these fluctuations could be attributed almost exclusively to noise. Therefore, the model is expected to be described with a coarser resolution.

If we employ the already mentioned reduced basis $\mathbf{B}$, filtering arises naturally from the construction of the reduced basis.

21.6.5 Nonlinear Models

Conceptually, considering nonlinear models does not change anything, but the fact that the model, $\mathbf{K}$ and the vectors it involves, must be samples from a small neighborhood of the data $\mathbf{U}$ or $\mathbf{F}$. By the manifold hypothesis, a dense enough sampling will behave linearly within this neighborhood.

Therefore, for each new datum $\mathbf{U}$, its corresponding cluster, κ, must be first identified. Then, the low-rank model at that cluster, $\mathbf{K}_\kappa^{LR}$, is chosen and the solution is evaluated according to $\mathbf{F} = \mathbf{K}_\kappa^{LR}\mathbf{U}$.

References

1. C. Argerich, R. Ibanez, A. Barasinski, F. Chinesta, Code2vect: An efficient heterogenous data classifier and nonlinear regression technique. C. R. Mecanique **347**, 754–761 (2019)
2. S. Brunton, J.L. Proctor, N. Kutz, Discovering governing equations from data by sparse identification of nonlinear dynamical systems. PNAS **113**(15), 3932–3937 (2016)

References

3. R. Ibanez, E. Abisset-Chavanne, A. Ammar, D. Gonzalez, E. Cueto, A. Huerta, J.L. Duval, F. Chinesta, A multi-dimensional data-driven sparse identification technique: the sparse Proper Generalized Decomposition. Complexity, Article ID 5608286 (2018)
4. B. Lusch, J.N. Kutz, S.L. Brunton, Deep learning for universal linear embeddings of nonlinear dynamics. Nature Commun. **9**, 4950 (2018)
5. P.J. Schmid, Dynamic mode decomposition of numerical and experimental data. J. Fluid Mech. **656**, 528 (2010)
6. M.O. Williams, G. Kevrekidis, C.W. Rowley, A data-driven approximation of the Koopman operator: extending dynamic mode decomposition. J. Nonlinear Sci. **25**(6), 1307–1346 (2015)
7. A. Reille, N. Hascoet, C. Ghnatios, A. Ammar, E. Cueto, J.L. Duval, F. Chinesta, R. Keunings, Incremental dynamic mode decomposition: a reduced-model learner operating at the low-data limit. C. R. Mecanique **347**, 780–792 (2019)

Open Access This chapter is licensed under the terms of the Creative Commons Attribution-NonCommercial-NoDerivatives 4.0 International License (http://creativecommons.org/licenses/by-nc-nd/4.0/), which permits any noncommercial use, sharing, distribution and reproduction in any medium or format, as long as you give appropriate credit to the original author(s) and the source, provide a link to the Creative Commons license and indicate if you modified the licensed material. You do not have permission under this license to share adapted material derived from this chapter or parts of it.

The images or other third party material in this chapter are included in the chapter's Creative Commons license, unless indicated otherwise in a credit line to the material. If material is not included in the chapter's Creative Commons license and your intended use is not permitted by statutory regulation or exceeds the permitted use, you will need to obtain permission directly from the copyright holder.

Part III
Around Reduction

Chapter 22
From Discretization to Model Order Reduction

To approximate the solution of problems usually encountered in sciences and engineering, defined in space and time, $u(x, t)$, a natural choice is to express it as a combination of a reduced number of functions with a physically or mathematically sound. This is at the origin of the Ritz method, that by considering functions $\mathcal{G}_i(\mathbf{x})$, $i = 1, \ldots, G$, the approximate solution reads

$$u(\mathbf{x}, t) \approx \sum_{i=1}^{G} \alpha_i(t) \mathcal{G}_i(\mathbf{x}),$$

where we introduce some time-dependent coefficients $\alpha_i(t)$. Their value is found by resorting to a projection method, such as the Galerkin method, for instance.

Determining the precise form of these functions is often not easy, however. An alternative approach consists of employing a general-purpose approximation basis. Polynomials (Lagrange, Chebyshev, Legendre, Fourier, ...) are the ubiquitous choice. Noting them by $\mathcal{P}_i(\mathbf{x})$, $i = 1, \ldots, Q$, the approximation now reads

$$u(\mathbf{x}, t) \approx \sum_{i=1}^{Q} \beta_i(t) \mathcal{P}_i(\mathbf{x}).$$

To guarantee an accurate approximation of our solution, a large enough number of these polynomials must be included in our approximation. This implies very often an increasing number of terms in the finite sum.

This approach is problematic if we deal with physical domains with complex geometry, due to the global character of the approximation considered so far. This motivated the substitution of global functions $\mathcal{P}_i(\mathbf{x})$ by compactly-supported piecewise polynomial functions $\mathcal{N}_i(\mathbf{x})$ associated to the N_n nodes supporting the approximation

$$u(\mathbf{x}, t) \approx \sum_{i=1}^{N_n} U_i(t)\mathcal{N}_i(\mathbf{x}),$$

with $U_i(t)$ the approximated solution at node $\mathbf{x}_i$, at time t. This approximation represents the finite element method, FEM, foundation.

While this approach was tremendously successful in the last decades, it becomes very computationally demanding for cases of industrial interest, needing too many nodes and time steps.

By using that approximate, the discretization of a PDE, assumed linear for the sake of simplicity, leads to the linear system

$$\mathbf{KU} = \mathbf{F}.$$

Here, matrix $\mathbf{K}$ represents the discrete form of the model, whereas vectors $\mathbf{U}$ and $\mathbf{F}$ contain the nodal unknowns and forcing terms (loads), respectively. Their sizes, when considering the finite element approximation described above are, respectively $N_n \times N_n, N_n \times 1$ and $N_n \times 1$.

22.1 Reduced Order Modelling

In many applications $\mathbf{U}$ is often defined in a subspace of dimension R, with R $\ll$ N_n [1]. Proper Orthogonal Decomposition, POD, embeds the solution onto a subspace of a small dimensionality R.

To obtain this subspace, POD employs a set of snapshots of the solution $\mathbf{U}_1, \ldots, \mathbf{U}_M$, obtained by the application of loads $\mathbf{F}_1, \ldots, \mathbf{F}_M$. In this way, a reduced-order basis $\boldsymbol{\phi}_i(\mathbf{x})$, $i = 1, \ldots, R$, with R $\ll N_n$ is constructed, and by using it, the approximation becomes

$$\mathbf{U}(t) \approx \sum_{i=1}^{R} \gamma_i(t)\boldsymbol{\phi}_i = \mathbf{B}\boldsymbol{\gamma}.$$

If we now introduce this reduced-order approximation into the discrete problem $\mathbf{KU} = \mathbf{F}$, and premultiplying by the transpose of $\mathbf{B}$ (equivalent to a Galerkin projection in the reduced basis), it results

$$(\mathbf{B}^T\mathbf{KB})\boldsymbol{\gamma} = \mathbf{B}^T\mathbf{F},$$

where the respective sizes of these elements of this reduced-order model are R $\times$ R, R $\times$ 1 and R $\times$ 1, respectively.

This procedure needs for some snapshots of the system to extract the reduced basis. However, it is possible to perform a simultaneous search of space and time functions, as the proper generalized decomposition, PGD, performs. We consider the generic approximation

$$u(\mathbf{x}, t) \approx \sum_{i=1}^{N} T_i(t) X_i(\mathbf{x}),$$

where both sequences of functions, T_i and X_i, are obtained by substituting this approximation into the weak form of the problem and then using a rank-one (greedy) enrichment to incrementally construct the separated representation.

22.2 More on PGD-Based Separated Representations

The main characteristic of PGD is that it constructs the approximation basis while it solves the problem, at the same time. For each problem an associated basis exists in which the solution is expressed.

Only a few terms in its approximation could be considered, thus leading to a reduced representation. On the contrary, in the limit, all the terms needed to approximate the solution up to a certain accuracy level could equally be considered.

We consider a general space-time separated representation

$$u(x, t) \approx \sum_{i=1}^{N} X_i(x) \, T_i(t), \tag{22.1}$$

where neither the time functions $T_i(t)$ nor the space functions $X_i(x)$ are *a priori* known. On the contrary, they will be computed on the fly when solving the problem.

One could make a step forward and assume that the solution of a multidimensional problem $u(x_1, \ldots, x_d)$ (as encountered in quantum chemistry, in the kinetic theory of complex fluids, or in statistical mechanics, ...) could be found in the separated form

$$u(x_1, x_2, \ldots, x_d) \approx \sum_{i=1}^{N} X_i^1(x_1) \, X_i^2(x_1) \cdots X_i^d(x_d),$$

and even more, expressing the 3D solution $u(x, y, z)$ as a finite sum decomposition involving lower dimensional functions

$$u(x, y, z) \approx \sum_{i=1}^{N} X_i(x) \, Y_i(y) \, Z_i(z),$$

or

$$u(x, y, z) \approx \sum_{i=1}^{N} X_i(x, y) \, Z_i(z),$$

and the solution of a parametric problem $u(\mathbf{x}, t, \mu_1, \ldots, \mu_P)$ as

$$u(\mathbf{x}, t, \mu_1, \ldots, \mu_P) \approx \sum_{i=1}^{N} X_i(\mathbf{x}) \, T_i(t) \prod_{k=1}^{P} M_i^k(\mu_k).$$

The performance of these separated representations is frequently very high, and the algorithms to be used to calculate the involved functions will be addressed in Chap. 24.

Reference

1. F. Chinesta, A. Huerta, G. Rozza, K. Willcox, Model order reduction, in *The Encyclopedia of Computational Mechanics, Second Edition, Erwin Stein, Rene de Borst, Tom Hughes Edt.*, John Wiley & Sons, Ltd. (2015)

Open Access This chapter is licensed under the terms of the Creative Commons Attribution-NonCommercial-NoDerivatives 4.0 International License (http://creativecommons.org/licenses/by-nc-nd/4.0/), which permits any noncommercial use, sharing, distribution and reproduction in any medium or format, as long as you give appropriate credit to the original author(s) and the source, provide a link to the Creative Commons license and indicate if you modified the licensed material. You do not have permission under this license to share adapted material derived from this chapter or parts of it.

The images or other third party material in this chapter are included in the chapter's Creative Commons license, unless indicated otherwise in a credit line to the material. If material is not included in the chapter's Creative Commons license and your intended use is not permitted by statutory regulation or exceeds the permitted use, you will need to obtain permission directly from the copyright holder.

Chapter 23
Proper Orthogonal Decomposition and Reduced Basis

23.1 Proper Orthogonal Decomposition Based Model Order Reduction

To introduce the Proper Orthogonal Decomposition, POD, let us begin by assuming that some solutions of the unknown field of interest $u(\mathbf{x}, t)$ are known at particular nodal positions $\mathbf{x}_i$ at discrete time instants $t_m = m \Delta t$, with $i \in [1, \ldots, N_n]$ and $m \in [1, \ldots, M]$. From now on, we denote $u(\mathbf{x}_i, t_m) \equiv u^m(\mathbf{x}_i) \equiv u_i^m$ and define $\mathbf{u}^m$ as the vector of nodal values u_i^m at time t_m.

In this framework, POD tries to obtain the most characteristic structure, from a statistical point of view, $\phi(\mathbf{x})$ among these $u^m(\mathbf{x})$, $\forall m$. For this purpose, one could apply the PCA introduced in Chap. 3. Here, we describe an equivalent derivation procedure based in the maximization of the scalar quantity

$$\lambda = \frac{\sum_{m=1}^{M} \left[\sum_{i=1}^{N_n} \phi(\mathbf{x}_i) u^m(\mathbf{x}_i) \right]^2}{\sum_{i=1}^{N_n} (\phi(\mathbf{x}_i))^2},$$

which is known to be equivalent to the eigenvalue problem $\mathbf{C}\phi = \lambda\phi$, with $\mathbf{C} = \sum_{m=1}^{M} \mathbf{u}^m (\mathbf{u}^m)^T$.

Once the eigenvalue problem is solved, we select the R eigenvectors ϕ_i associated with the highest eigenvalues to construct a reduced approximation basis for our original problem. The advantage comes from the fact that, in practice, $R \ll N_n$ very often.

We place these R selected eigenvectors in the columns of matrix $\mathbf{B}$, and then $\mathbf{U}$ is expressed in that reduced basis according to $\mathbf{U} = \mathbf{B}\boldsymbol{\gamma}$. Then, as previously discussed, the system $\mathbf{KU} = \mathbf{F}$ becomes $\mathbf{k}\boldsymbol{\gamma} = \mathbf{f}$, by premultiplying by $\mathbf{B}^T$ and using the reduced approximation of $\mathbf{U}$.

An important drawback of this procedure is related to the size of the eigenproblem to solve. In fact, the size of the correlation matrix, $\mathbf{C}$, is $N_n \times N_n$, with N_n scaling with the number of nodes involved in the problem discretization. This number can reach in some applications millions or even more. The so-called Snapshot-POD allows alleviating the just referred issue.

23.2 Snapshot-POD

If the solutions snapshots $\mathbf{u}^m$ are placed in the columns of matrix $\mathbf{U}$, the correlation matrix considered by the POD just described results $\mathbf{C} = \mathbf{U}\mathbf{U}^T$.

The so-called Schmidt-Hilbert formulation [1, 2] assumes that each eigenvector ϕ_i can be expressed as a linear combination of the snapshots, that is

$$\phi_i = \sum_{k=1}^{M} \alpha_k^i \mathbf{u}^k = \mathbf{U}\alpha,$$

that replaced into $\mathbf{U}\mathbf{U}^T\phi = \lambda\phi$, results in $\mathbf{U}\mathbf{U}^T\mathbf{U}\alpha = \lambda\mathbf{U}\alpha$, or $\mathbf{U}\left((\mathbf{U}^T\mathbf{U})\alpha - \lambda\alpha\right) = \mathbf{0}$, from which, it results

$$(\mathbf{U}^T\mathbf{U})\alpha = \lambda\alpha,$$

that defines another eigenproblem, but this time of size $M \times M$, much smaller than its standard POD counterpart. From the resulting α_i, the eigenmode ϕ_i results $\phi_i = \mathbf{U}\alpha_i, \forall i$.

23.3 Hyper-Reduction

When using the FEM the approximation induces the size of the coefficients matrix $\mathbf{K}$. When using the reduced basis, only R coefficients must be computed, and at first view calculating a model of size N_n to compute at the end only R values could seem disproportionate.

Hyper-reduction [3, 4] consists in considering only a few rows of $\mathbf{K}$, its number scaling with R, in fact slightly higher than R, to define $\mathbf{K}_r$ to derive the reduced model $\mathbf{k}_r$. This strategy can be viewed as a sort of Petrov-Galerkin technique, where the test functions select the rows to be extracted from $\mathbf{K}$.

This procedure is particularly appealing in the nonlinear case where the coefficient matrix must be calculated many times, being its main drawback the necessity of having a strategy for extracting the most valuable rows of matrix $\mathbf{K}$.

23.4 Compressed Sensing Based Adaptive Mode Selection

In the case of transient nonlinear models, the choice of the POD modes to be selected to define the reduced basis is a tricky issue, because that reduced basis could evolve and the modes to be retained can differ and evolve.

Instead of enforcing a truncation to the first modes of the POD decomposition, one possibility consists in retaining all them, and using compressed sensing to select the ones to be considered in the solution approximation by enforcing sparsity.

23.5 Reduced Basis, RB

Reduced basis method [5] proceeds by computing different snapshots for different parameters choices $\mathbf{u}_i \equiv \mathbf{u}(\mu_i)$, sequentially in order to construct a hierarchical approximation. Then, instead of applying the POD to recover an orthonormal basis, RB uses a simple orthogonalization (e.g. Gram-Schmidt), and use the resulting basis for projecting the solution within a Galerkin framework.

The main difference of RB (with respect to standard POD) is the sampling procedure that in the case of RB is driven by an error estimator with solid mathematical foundations. Thus, if Ξ defines a finite sample of points in the parametric space (generated for example by using a Monte-Carlo procedure), after being computed the solution $\mathbf{u}$ at $n-1$ parameters choices, at the present iteration we select the new point (new value of the model parameters) as

$$\mu^n = \arg\max_{\mu \in \Xi} \Delta_{n-1}(\mu),$$

where $\Delta_n(\mu)$ is a sharp, asymptotically inexpensive a posteriori error bound for an appropriate norm $\|\mathbf{u}(\mu) - \mathbf{u}^n(\mu)\|$.

23.6 Gappy-POD

The Gappy-POD [6, 7] applies when the different snapshots on which the POD applies are incomplete. As discussed in the original works, the snapshots are generated by applying random masks $\mathbf{m}_n$ on the highly-resolved snapshots $\mathbf{u}_k$, generating a number of *marred* snapshots $\mathbf{u}_k^m$ (different masks could apply on the same highly resolved snapshot), where the index $\bullet^m$ refers to the fact that a mask applied on it. Each marred snapshot $\mathbf{u}_k^m$ takes zero values in the locations in which the applied mask operates.

To apply the POD, the marred snapshots must be corrected. The simplest possibility is considering at the masked points (containing zero values) the average at that position of all the marred snapshots, assuming that there is some snapshots in which the solution at that point is not zero. Thus each $\mathbf{u}_k^m$ gives its corrected counterpart $\hat{\mathbf{u}}_k^m$.

Then, standard POD applies to compute the reduced basis $\mathbf{B}$, on which each corrected snapshot in projected (considering only the proper data, and then excluding in the least squares projection the corrected values). Thus, we will obtain vectors γ_k that approximate masked snapshots $\mathbf{u}_k^m$, and then a better correction is obtained from $\hat{\mathbf{u}}_k^m = \mathbf{B}\boldsymbol{\gamma}_k$. Then the POD can be calculated again and the corrected snapshots recomputed until reaching convergence (fixed point iteration).

23.7 Nonlinear Models

When a model is nonlinear, after computing the reduced solution from $\boldsymbol{\gamma}$, the solution must be reconstructed $\mathbf{U} = \mathbf{B}\boldsymbol{\gamma}$, and now, from the nodal values of the unknown field, the last can be evaluated everywhere and then the model $\mathbf{K}$ (or at least its nonlinear part) reconstructed, i.e. $\mathbf{K}(\mathbf{U}(\boldsymbol{\gamma}))$.

To alleviate the computational cost, one route consists in using the just described hyper-reduction. Others explore the same idea, the one of reducing the evaluation cost of the nonlinear term. Gappy-POD [6] (just addressed) proceeds in this way, and is one of the components of the so-called GNAT (Gauss-Newton with Approximated Tensors) [8], and the so-called Missing Point Estimation [9] reduces the cost by computing Galerkin projections over a restricted subset of the spatial domain.

Another very appealing route consists in approximating or interpolating the nonlinear term, as described later.

23.7.1 Nonlinearities Interpolated: The Discrete Empirical Interpolation Method, DEIM

If we consider a generic nonlinear function $g(u(\mathbf{x}, t))$, its evaluation in the whole state space may be expensive, and even compromise the efficiency of reduced model techniques.

The Empirical Interpolation Method, EIM, and its discrete counterpart, DEIM, [10, 11], use the POD just described to approximate the nonlinear term in the resulting reduced-order basis as

$$g(\mathbf{x}, t) \approx \sum_{n=1}^{N_g} a_g^n(t) \phi_g^n(\mathbf{x}).$$

In order to compute the approximation coefficients a_g^n, with $n = 1, \ldots, N_g$, the nonlinear function is reconstructed at any time t_j at only N_g nodes, denoted by $\mathbf{x}_g^i$, with $i = 1, \ldots, N_g$. It results in the following linear system to be solved at each time:

$$\begin{pmatrix} g(\mathbf{x}_g^1, t_j) \\ \vdots \\ g(\mathbf{x}_g^{N_g}, t_j) \end{pmatrix} = \begin{pmatrix} \phi_g^1(\mathbf{x}_g^1) & \cdots & \phi_g^{N_g}(\mathbf{x}_g^1) \\ \vdots & \ddots & \vdots \\ \phi_g^1(\mathbf{x}_g^{N_g}) & \cdots & \phi_g^{N_g}(\mathbf{x}_g^{N_g}) \end{pmatrix} \begin{pmatrix} a_g^1(t_j) \\ \vdots \\ a_g^{N_g}(t_j) \end{pmatrix}.$$

The choice of the points $\mathbf{x}_g^i, i = 1, \ldots, N_g$ deserves some comments. These points are a priori arbitrary, the only constraint being for the previous matrix to be invertible. However, these points can be obtained using a greedy strategy in order to place them so as to capture as much information as possible. The resulting points are often called *magic points*. Their computation is described below.

The procedure starts by considering $\mathbf{x}_g^1 = \mathrm{argmax}_{\mathbf{x}} |\phi_g^1(\mathbf{x})|$. Then d_1 is computed from $d_1 \phi_g^1(\mathbf{x}_g^1) = \phi_g^2(\mathbf{x}_g^1)$, that allows defining the residual $r_2(\mathbf{x})$, i.e. a contribution in $\phi_g^2(\mathbf{x})$ that cannot be explained by $\phi_g^1(\mathbf{x})$, from which computing point $\mathbf{x}_g^2$:

$$\mathbf{x}_g^2 = \mathrm{argmax}_{\mathbf{x}} |r_2(\mathbf{x})| \quad \text{with} \quad r_2(\mathbf{x}) = \phi_g^2(\mathbf{x}) - d_1 \phi_g^1(\mathbf{x}).$$

As by construction $r_2(\mathbf{x}_g^1) = 0$ we can ensure $\mathbf{x}_g^2 \neq \mathbf{x}_g^1$.

The procedure is generalized to obtain the other points involved in the interpolation procedure. Thus, to obtain point $\mathbf{x}_g^i$ we consider

$$\mathbf{x}_g^i = \mathrm{argmax}_{\mathbf{x}} |r_i(\mathbf{x})| \quad \text{with} \quad r_i(\mathbf{x}) = \phi_g^i(\mathbf{x}) - \sum_{j=1}^{i-1} d_j \phi_g^j(\mathbf{x}).$$

Coefficients $d_1, \ldots, d_{i-1}$ must be chosen so as to ensure that $\mathbf{x}_g^i \neq \mathbf{x}_g^j, \forall j < i$. To this end, we enforce the residual $r_i(\mathbf{x})$ to vanish at each location $\mathbf{x}_g^j$, with $j < i$ by solving:

$$r_i(\mathbf{x}_g^j) = 0 = \phi_g^i(\mathbf{x}_g^j) - \sum_{l=1}^{i-1} d_l \phi_g^l(\mathbf{x}_g^j), \quad j = 1, \ldots, i-1,$$

that constitutes a linear system whose solution leads to the coefficients $d_1, \ldots, d_{i-1}$.

23.7.2 Trajectory Piece-Wise Linear, TPWL

Model Order Reduction of nonlinear dynamical systems remains challenging due to the necessity of evaluating the nonlinear term at each iteration, as just discussed.

If we consider the dynamical problem

$$\begin{cases} \dot{\mathbf{x}}(t) = \mathbf{f}(\mathbf{x}(t)) + \mathbf{B}\mathbf{u}(t) \\ \mathbf{y}(t) = \mathbf{C}^T \mathbf{x}(t), \end{cases}$$

with $\mathbf{x}(t) \in \mathbb{R}^N$, $\mathbf{f} : \mathbb{R}^N \to \mathbb{R}^N$ a nonlinear function, $\mathbf{B}$ the N × M input matrix, $\mathbf{u}(t) \in \mathbb{R}^M$ the input signal, $\mathbf{C}$ the N × K output matrix, and $\mathbf{y}(t) \in \mathbb{R}^K$ the output signal (e.g. the measure).

Model Order Reduction techniques extract and then employ a reduced approximation basis that allows projecting any state into that reduced basis of dimension R, with $R \ll N$. The projection reads

$$\mathbf{x} = \mathbf{Vz},$$

that allows reducing the dynamical system

$$\begin{cases} \mathbf{V\dot{z}}(t) = \mathbf{f}(\mathbf{Vz}(t)) + \mathbf{Bu}(t) \\ \mathbf{y}(t) = \mathbf{C}^T \mathbf{Vz}(t), \end{cases}$$

that projected into the reduced basis, and taking into account its orthonormality, leads to

$$\begin{cases} \dot{\mathbf{z}}(t) = \mathbf{V}^T \mathbf{f}(\mathbf{Vz}(t)) + \mathbf{V}^T \mathbf{Bu}(t) \\ \mathbf{y}(t) = \mathbf{C}^T \mathbf{Vz}(t), \end{cases}$$

being the main difficulty of its use, the evaluation of $\mathbf{f}(\mathbf{Vz})$ at each time step [12, 13].

TPWL consider a simple linearization around a functioning point $\mathbf{x}_0$, expressed from

$$\mathbf{f}(\mathbf{x}) \approx \mathbf{f}(\mathbf{x}_0) + \mathbf{J}_0(\mathbf{x} - \mathbf{x}_0) + \frac{1}{2}\mathbf{H} : [(\mathbf{x} - \mathbf{x}_0) \otimes (\mathbf{x} - \mathbf{x}_0)],$$

where $\mathbf{J}_0$ is the second order Jacobian and $\mathbf{H}$ the third order Hessian, both at the functioning state $\mathbf{x}_0$, with $\otimes$ the tensor product, and : the tensor product twice contracted.

Restricting the previous expansion to its linear term, and introducing it into the reduced dynamical system, it results

$$\begin{cases} \dot{\mathbf{z}}(t) = \mathbf{V}^T \mathbf{f}(\mathbf{x}_0) + \mathbf{V}^T \mathbf{J}_0 \mathbf{Vz} - \mathbf{V}^T \mathbf{J}_0 \mathbf{x}_0 + \mathbf{V}^T \mathbf{Bu}(t) \\ \mathbf{y}(t) = \mathbf{C}^T \mathbf{Vz}(t), \end{cases}$$

whose main limitation is the cost when increasing the expansion order, needed for representing strong nonlinear behaviors as soon as the current state $\mathbf{x}$ moves far from the state that served for computing the expansion $\mathbf{x}_0$.

It is there where TPWS proposes generating a number, s, of linearized models around the states $\mathbf{x}_0, ..., \mathbf{x}_{s-1}$:

$$\dot{\mathbf{x}}(t) = \mathbf{f}(\mathbf{x}_i) + \mathbf{J}_i(\mathbf{x}(t) - \mathbf{x}_i) + \mathbf{Bu}(t),$$

where $\mathbf{x}_0$ is the initial state and $\mathbf{x}_i$ other states along one of the trajectories of the dynamical system extracted in the offline training stage.

The weighted combination of all those linearized forms reads

$$\dot{\mathbf{x}}(t) = \sum_{i=0}^{s-1} \omega_i(\mathbf{x})\mathbf{f}(\mathbf{x}_i) + \sum_{i=0}^{s-1} \omega_i(\mathbf{x})\mathbf{J}_i(\mathbf{x} - \mathbf{x}_i) + \mathbf{Bu},$$

with the weighs verifying

$$\sum_{i=1}^{s-1} \omega_i(\mathbf{x}) = 1,$$

and with the value of $\omega_i(\mathbf{x})$ decreasing when the distance between $\mathbf{x}$ and $\mathbf{x}_i$ increases.

Thus, its reduced counterpart reads:

$$\dot{\mathbf{z}}(t) = \sum_{i=0}^{s-1} \tilde{\omega}_i(\mathbf{z}) \mathbf{V}^T \{\mathbf{f}(\mathbf{x}_i) - \mathbf{J}_i \mathbf{x}_i\} + \sum_{i=0}^{s-1} \tilde{\omega}_i(\mathbf{z}) \mathbf{V}^T \mathbf{J}_i \mathbf{V} \mathbf{z} + \mathbf{V}^T \mathbf{B} \mathbf{u},$$

where now the weighing proceeds in the reduced space, with functions $\tilde{\omega}_i(\mathbf{z})$ having the same properties that the ones that were discussed on their counterparts $\omega_i(\mathbf{x})$.

The reduced basis can be extracted offline by using for instance the POD, on some trajectories of the dynamical system for different input data $\mathbf{u}(t)$. The linearization points must be chosen in order to cover up to a certain precision, the functioning domain, while ensuring that in between, the linearized reduced model remains convenable.

23.8 Grassman and Barycentric Interpolation

23.8.1 Grassman Manifold

This section makes use of the concepts introduced in Chap. 11, and reported in [14]. We suppose that two reduced bases $\mathbf{B}_0$ and $\mathbf{B}_1$ have been extracted from the transient solution of a given problem, for two different choices of the model parameter, here labelled by $s = 0$ and $s = 1$. Now, the key question is to provide a reduced basis $\mathbf{B}_s$ for any value of s ($0 < s < 1$).

First we proceed to compute $\dot{\mathbf{Y}}_0$ from the *logarithmic map* $\text{Log}_{S_0} S_1$ (where S_0 and S_1 denotes the points in the manifold)

$$(\mathbf{I} - \mathbf{B}_0 \mathbf{B}_0^T) \mathbf{B}_1 (\mathbf{B}_0^T \mathbf{B}_1)^{-1} = \mathbf{U} \boldsymbol{\Sigma} \mathbf{V}^T$$

with

$$\dot{\mathbf{Y}}_0 = \mathbf{U} \tan^{-1}(\boldsymbol{\Sigma}) \mathbf{V}^T.$$

Now $\mathbf{B}_1$ could be determined by the *exponential map*, according to

$$\mathbf{B}_1 = \mathbf{B}_0 \mathbf{V} \cos(\tanh^{-1}(\boldsymbol{\Sigma})) + \mathbf{U} \sin(\tan^{-1}(\boldsymbol{\Sigma})),$$

with the intermediate basis, associated with $0 < s < 1$,

$$\mathbf{B}_s = \mathbf{B}_0 \mathbf{V} \cos(s \tanh^{-1}(\mathbf{\Sigma})) + \mathbf{U} \sin(s \tan^{-1}(\mathbf{\Sigma})).$$

This procedure is generalizable for richer samplings and multi-parametric settings.

23.8.2 From Grassman to Barycentric Interpolation of Reduced Bases

Interpolating reduced bases in the tangent space was successfully performed by using the explicit expressions of the exponential and logarithmic mappings just reported.

Here, using the notation employed in [15], we consider N_p reduced bases of dimension R, much smaller than the dimension N of the embedding space $\mathbb{R}^N$, $\mathbf{\Phi}_k$, $k = 1, ..., N_p$, associated with N_p choices of the parameter μ: $\mu_1, ..., \mu_{N_p}$.

As soon as the tangent space to the manifold, for instance at $\mathbf{\Phi}_0$, is constrcuted, the so-called logarithmic map allows us to compute matrices $\mathbf{\Psi}_k$ in the tangent space, representing the initial velocity of the geodesic path starting at $\mathbf{\Phi}_0$ and finishing at $\mathbf{\Phi}_k$.

Now, for a given value of the parameter μ, matrices $\mathbf{\Psi}_k$ can be interpolated to give $\mathbf{\Psi}$. Then, the use of the so-called exponential map will lead to the searched reduced basis $\mathbf{\Phi}$, to be employed, for instance, in a Galerkin projection, for computing the problem solution associated with the value μ of the parameter. The main issue of such a procedure, based on the Grassmannian manifold, is the dependence on the chosen tangent space.

Another way of defining an interpolator able to perform on a Riemannian manifold, avoiding the use of a tangent reference space, is the one based on the use of the Riemannian barycenter [15].

The barycentric interpolation looks for the reduced basis $\mathbf{\Phi}$ (related to the value of the parameter μ) minimizing $\mathcal{C}(\mathbf{\Phi})$:

$$\mathcal{C}(\mathbf{\Phi}) = \frac{1}{2} \sum_{k=1}^{N_p} \omega_k(\mu) \text{dist}^2(\mathbf{\Phi}_k, \mathbf{\Phi}), \tag{23.1}$$

with in general the weighting functions verifying the usual conditions:

$$\begin{cases} \sum_{k=1}^{N_p} \omega_k(\mu) = 1 \\ \omega_k(\mu_l) = \delta_{kl} \end{cases},$$

where δ is the Kroenercker delta. For that purpose many choices exist, among them, the use of Lagrangian functions, radial basis, ...

The so-called weighted Karcher barycenter generalizes the classical weighted mean in a vector space to Riemannian manifolds. In [15], the authors proposed to construct a sequence of bases $\mathbf{\Phi}^m$ that converges to the searched reduced basis $\mathbf{\Phi}$, to alleviate the difficulty of ensuring the existence and uniqueness of a global minimum of (23.1). The Riemannian gradient descent was employed. Thus, following [15]:

- Given $\mathbf{\Phi}^m$, the gradient of Eq. (23.1) is calculated

$$\nabla \mathcal{C}(\mathbf{\Phi}^m) = -\sum_{k=1}^{N_p} \omega_k(\mu) \log_{\mathbf{\Phi}^m} \mathbf{\Phi}_k,$$

where the geodesic logarithm $\log_{\mathbf{\Phi}^m} \mathbf{\Phi}_k$ results from the SVD decomposition $\mathbf{\Phi}^{mT} \mathbf{\Phi}_k = \mathbf{U}_k^m \mathbf{\Sigma}_k^m \mathbf{V}_k^m$, then $\log_{\mathbf{\Phi}^m} \mathbf{\Phi}_k = \mathbf{\Phi}_k \mathbf{Q}_k^m - \mathbf{\Phi}^m$, with $\mathbf{Q}_k^m = \mathbf{V}_k^m \mathbf{U}_k^{mT}$.

- If the norm of the gradient is not small enough, a correction is applied according to

$$\mathbf{\Phi}^{m+1} = \mathrm{Exp}_{\mathbf{\Phi}^m}(-\nabla \mathcal{C}(\mathbf{\Phi}^m)).$$

Thus, the updating results

$$\mathbf{\Phi}^{m+1} = \sum_{k=1}^{N_p} \omega_k(\mu) \mathbf{\Phi}_k \mathbf{Q}_k^m.$$

References

1. L. Sirovich, Turbulence and the dynamics of coherent structures. Quart. J. Appl. Math. **45**, 561–590 (1987)
2. H.M. Park, D.H. Cho, Low dimensional modeling of flow reactors. Int. J. Heat Mass Transfer **39**(16), 3311–3323 (1996)
3. D. Ryckelynck, A priori hyperreduction method: an adaptive approach. J. Comput. Phys. **202**, 346–366 (2005)
4. T. Chapman, P. Avery, P. Collins, C. Farhat, Accelerated mesh sampling for the hyper reduction of nonlinear computational models. Int. J. Num. Meth. Engrg. **109**(12), 1623–1654 (2017)
5. F. Chinesta, A. Huerta, G. Rozza, K. Willcox, Model order reductio, in *the Encyclopedia of Computational Mechanics, Second Edition, Erwin Stein, Rene de Borst, Tom Hughes Edt.*, John Wiley & Sons, Ltd. (2015)
6. R. Everson, L. Sirovich, Karhunen-Loeve procedure for gappy data. J. Opt. Soc. Am. A **12**(8), 1657–1664 (1995)
7. H. Gunes, S. Sirisup, G.E. Karniadakis, Gappy data: to Krig or not to Krig? J. Comput. Phys. **212**, 358–382 (2006)
8. K. Carlberg, C. Bou-Mosleh, C. Farhat, Efficient non-linear model reduction via a least-squares Petrov-Galerkin projection and compressive tensor approximations. Int. J. Numer. Methods Eng. **86**, 155–181 (2011)
9. P. Astrid, S. Weiland, K. Willcox, T. Backx, Missing point estimation in models described by proper orthogonal decomposition. IEEE Trans. Autom. Control **53**, 2237–2251 (2008)
10. M. Barrault, Y. Maday, N. Nguyen, A. Patera, An empirical interpolation method: application to efficient reduced-basis discretization of partial differential equations. C.R. Acad. Sci. I-Math. **339**(9), 667–672 (2004)
11. F. Chazal, B. Michel, An introduction to topological data analysis: fundamental and practical aspects for data scientists. Journal de la Societe Francaise de Statistiques (2017)

12. M. Rewienski, J. White, A trajectory piecewise-linear approach to model order reduction and fast simulation of nonlinear circuits and micromachined devices. In IEEE Trans. Comput. Aided Design Integra. Circuits Syst. **22**(2), 155–170 (2003)
13. M.H. Malik, D. Borzacchiello, F. Chinesta, P. Diez, Reduced order modeling for transient simulation of power systems using trajectory piece-wise linear approximation. Adv. Model. Simul. Eng. Sci. 3(31) (2016)
14. D. Amsallem, C. Farhat, Interpolation method for adapting reduced-order models and application to aeroelasticity. AIAA J. **46**(7), 1803–1813 (2008)
15. M. Oulghelou, C. Allery, R. Mosquere, Parametric reduced order models based on a Riemannian Barycentric interpolation. Int. J. Num. Meth. Eng. **122**(22), 6623–6640 (2021)

Open Access This chapter is licensed under the terms of the Creative Commons Attribution-NonCommercial-NoDerivatives 4.0 International License (http://creativecommons.org/licenses/by-nc-nd/4.0/), which permits any noncommercial use, sharing, distribution and reproduction in any medium or format, as long as you give appropriate credit to the original author(s) and the source, provide a link to the Creative Commons license and indicate if you modified the licensed material. You do not have permission under this license to share adapted material derived from this chapter or parts of it.

The images or other third party material in this chapter are included in the chapter's Creative Commons license, unless indicated otherwise in a credit line to the material. If material is not included in the chapter's Creative Commons license and your intended use is not permitted by statutory regulation or exceeds the permitted use, you will need to obtain permission directly from the copyright holder.

Chapter 24
The Proper Generalized Decomposition

This section addresses the original, intrusive, version of the PGD solver, before considering later its non-intrusive counterparts. The PGD was proposed for solving problems defined in high-dimensional spaces, as encountered in quantum chemistry or statistical mechanics [1, 2], as well as for separating space and time to obtain non-incremental integrators [3–6].

Later, it was extended for separating the space coordinates [7, 8]. A tensor form of it was given in [9], that facilitated its computational implementation. In [10] it was reformulated to define a transfer function, allowing very efficient integrations of structural dynamics problems. *A posteriori* error estimators and indicators were considered in [11–13].

One of the most appealing applications of the PGD was the solution of multi-parametric problems [14], where parameters can also concern the geometry itself [15].

Domain decomposition was considered in [16, 17] and its coupling with finite elements in [18]. Different issues related to advective stabilization, nonlinearities, non-affine decompositions, instabilities and bifurcations, ... were addressed in [19–22].

The interested reader can refer to some recent reviews [23–25] as well as to the numerous references therein.

24.1 Extended Separated Representations

First, we consider the parametric heat transfer equation

$$\frac{\partial u}{\partial t} - \kappa \Delta u - f = 0,$$

with homogeneous initial and boundary conditions. A completely new approach to the problem arises by simply considering the conductivity k as a new coordinate, which will be defined in its own interval of interest Ω_k. Thus, $(\mathbf{x}, t, k) \in \Omega_x \times \Omega_t \times \Omega_k$.

The associated parametric problem weak form reads

$$\int_{\Omega_x \times \Omega_t \times \Omega_k} u^* \left(\frac{\partial u}{\partial t} - \kappa \Delta u - f \right) d\mathbf{x}\, dt\, dk = 0,$$

with the solution searched in the separated form

$$u(\mathbf{x}, t, k) \approx \sum_{i=1}^{N} X_i(\mathbf{x}) T_i(t) K_i(k).$$

At one arbitrary iteration $n < N$ the solution $u^n(\mathbf{x}, t, k)$ is assumed to be expressible in the form

$$u^n(\mathbf{x}, t, k) = \sum_{i=1}^{n} X_i(\mathbf{x}) T_i(t) K_i(k),$$

so that an enrichment of this approximation, $u^{n+1}(\mathbf{x}, t, k)$, will be

$$u^{n+1}(\mathbf{x}, t, k) = u^n(\mathbf{x}, t, k) + X_{n+1}(\mathbf{x}) T_{n+1}(t) K_{n+1}(k).$$

The test function u^* for this approximation could be

$$u^*(\mathbf{x}, t, k) = X^*(\mathbf{x}) T_{n+1}(t) K_{n+1}(k) + X_{n+1}(\mathbf{x}) T^*(t) K_{n+1}(k) + X_{n+1}(\mathbf{x}) T_{n+1}(t) K^*(k).$$

As usual, trial and test functions are substituted into the extended weak form. After an appropriate linearization, finite element approximations to functions $X_{n+1}(\mathbf{x})$, $T_{n+1}(t)$ and $K_{n+1}(k)$ are found. The easiest way of linearizing the problem is by an alternated directions, fixed point algorithm. It proceeds through the following steps:

- Arbitrarily initialize at the first iteration $T^0_{n+1}(t)$ and $K^0_{n+1}(k)$.
- With $T^{p-1}_{n+1}(t)$ and $K^{p-1}_{n+1}(k)$ given at the previous, $p-1$, iteration of the non linear solver, all the integrals in $\Omega_t \times \Omega_k$ are computed, leading to a boundary value problem for $X^p_{n+1}(\mathbf{x})$.
- With $X^p_{n+1}(\mathbf{x})$ just computed and $K^{p-1}_{n+1}(k)$ given at the previous iteration of the nonlinear solver, all the integrals in $\Omega_x \times \Omega_k$ are computed, leading to an one-dimensional initial value problem for $T^p_{n+1}(t)$.
- With $X^p_{n+1}(\mathbf{x})$ and $T^p_{n+1}(t)$ just updated, all the integrals in $\Omega_x \times \Omega_t$ are performed, leading to an algebraic problem for $K^p_{n+1}(k)$.

24.2 Nonlinear Models

Since PGD is based on variable separation, addressing nonlinear models requires performing a separated representation of all the nonlinear terms to recover an affine problem structure. There are many possibilities to perform such an affine representation.

One possibility consists in reconstructing the parametric field of interest $f(u(\mathbf{x}, t, p))$ (with p the model parameter) at each $\mathbf{x}_i$, t_j and p_k, leading to the multi-dimensional tensor f_{ijk}. Then, it can be separated by using the SVD (in the 2D case) or its high-dimensional counterpart, the HOSVD, when addressing more than two dimensions, is very close to the PGD procedure to separate functions.

Another possibility consists in using the cross decomposition. However, all the just referred techniques fail when addressing many dimensions. In those cases, sparse regressions seem an appealing route to proceed in highly-dimensional spaces.

In what follows we consider a model involving the square of the unknown field $u(\mathbf{x}, t)$, i.e. u^2

$$\frac{\partial u}{\partial t} - \kappa \Delta u + u^2 = f(\mathbf{x}, t).$$

The separated representation of the problem solution reads

$$u(\mathbf{x}, t) \approx \sum_{i=1}^{N} X_i(\mathbf{x}) T_i(t).$$

We assume that at iteration $n-1$, with $n \leq N$, the $n-1$ first modes (X_i, T_i), $i = 1, \ldots, n-1$, are already known and that at present iteration we search the new enrichment functional product $X_n(\mathbf{x}) T_n(t)$, as the fixed point of the iteration that computes X_n^q and T_n^q, where the super-index refers to the non-linear fixed point algorithm.

When the problem involves nonlinearities, we considered in our former works different approaches, some of them reported below.

24.2.1 Incremental Linearization

The simplest strategy to compute the non-linear term u^2 is to obtain it from the solution at the previous enrichment iteration, that is, from u^{n-1}

$$u^2 \approx \left(\sum_{i=1}^{n-1} X_i(\mathbf{x}) T_i(t) \right)^2,$$

and then to express it in a separated form.

24.2.2 Newton Linearization

From the solution at iteration $n-1$, u^{n-1}, the solution at the next iteration can be written as $u^n = u^{n-1} + \tilde{u}$ where $\tilde{u}$ is the solution of the linearized problem

$$\frac{\partial \tilde{u}}{\partial t} - \kappa \Delta \tilde{u} + 2 u^{n-1} \tilde{u} = -\mathcal{R}(u^{n-1}),$$

with $\mathcal{R}(u^{n-1})$ the problem residual associated with $u^{n-1}(\mathbf{x}, t)$.

24.2.3 Enhancing Iteration Procedures

While both procedures are shown to converge in practice, no significant differences in the number of required iterations were obtained. Both the convergence rate and the computing time were similar. However, we found that even in some cases for which the exact solution can be represented by a single functional product, i.e. $u^{ex}(\mathbf{x}, t) = X^{ex}(\mathbf{x}) T^{ex}(t)$ the non linear solver produces a solution composed of many sums

$$u(\mathbf{x}, t) \approx \sum_{i=1}^{N} X_i(\mathbf{x}) T_i(t),$$

with $N > 1$. The main reason for that is that the number of sums is in this case subsidiary of the convergence rate of the non-linear solver.

An improved fixed point strategy can be obtained by approximating the nonlinear term by using

$$u^{n-1} + X_n^{q-1}(\mathbf{x}) \cdot T_n^{q-1}(t),$$

that for a solution like $u^{ex}(\mathbf{x}, t) = X^{ex}(\mathbf{x}) T^{ex}(t)$, converges after computing the first functional couple. In that sense the solver is optimal.

The main difficulty related to the use of standard linearizations lies in the necessity of evaluating the non-linear term as previously indicated.

Other strategies were considered in our former works, in particular the Asymptotic Numerical Method, ANM, and the Discrete Empirical Interpolation Method, DEIM, revisited below.

24.2.4 Asymptotic Numerical Method, ANM

ANM proceeds, profiting of the polynomial nonlinearity (even if more complex non-linearities can be also addressed), by introducing the loading parameter λ affecting the non-linear term, i.e.

24.2 Nonlinear Models

$$\frac{\partial u}{\partial t} - \kappa \Delta u = -\lambda u^2 + f(\mathbf{x}, t) \tag{24.1}$$

We denote by u_0 the solution related to $\lambda = \lambda_0 = 0$ that can be computed easily because it corresponds to the solution of the linear problem. The searched solution is the one related to $\lambda = 1$. Now, we define an asymptotic expansion of the unknown field u as well as of the loading parameter λ by considering powers of the expansion parameter a:

$$\begin{cases} u = u_0 + a u_1 + a^2 u_2 + \ldots \\ \lambda = \lambda_0 + a \lambda_1 + a^2 \lambda_2 + \ldots \end{cases}$$

The non linear term can be written as

$$u^2 = (u^2)_0 + a(u^2)_1 + a^2(u^2)_2 + \cdots + a^p(u^2)_p + \ldots$$

where $(u^2)_p$ reads $(u^2)_p = \sum_{i=0}^{i=p} u_i u_{p-i} = 2 u_0 u_p + \sum_{i=1}^{i=p-1} u_i u_{p-i}$.

Introducing these expansions into the problem, and identifying the different powers of a, results in a sequence of problems, one at each order (power of a), all of them having the same differential operator, and whose right hand members depend on the solutions computed at lower orders.

In the case of higher powers, the introduction of new variables avoids the computation of powers higher than 2. More complex nonlinearities need a polynomial expansion prior to apply the ANM procedure just addressed.

24.2.5 Discrete Empirical Interpolation Method, DEIM

We consider the solution of $\mathcal{L}^L(u(\mathbf{x}, t)) = \mathcal{L}(u(\mathbf{x}, t)) + f(\mathbf{x}, t)$ where $\mathcal{L}^L(u(\mathbf{x}, t))$ and $\mathcal{L}(\bullet)$ are respectively a linear and a nonlinear operators.

We first address the PGD solution $u^0(\mathbf{x}, t)$ of the linear problem $\mathcal{L}^L(u(\mathbf{x}, t)) = f(\mathbf{x}, t)$, that reads

$$u^0(\mathbf{x}, t) \approx \sum_{i=1}^{i=N^0} X_i^0(\mathbf{x}) T_i^0(t),$$

that allows to define the reduced approximation basis $\mathcal{B}^0 = \{\tilde{X}_1^0 \tilde{T}_1^0, \ldots, \tilde{X}_{N^0}^0 \tilde{T}_{N^0}^0\}$ which contains the normalized functions: $\tilde{X}_i^0 = X_i^0 / \|X_i^0\|$ and $\tilde{T}_i^0 = T_i^0 / \|T_i^0\|$.

Now, we could define an interpolation of $\mathcal{L}(u)$ by using basis $\mathcal{B}^0$. For this purpose we consider N^0 points $(\mathbf{x}_j^0, t_j^0)$, $j = 1, \ldots, N^0$, and we enforce

$$\mathcal{L}(u^0(\mathbf{x}^0, t^0))\big|_{\mathbf{x}_j^0, t_j^0} = \sum_{i=1}^{i=N^0} \xi_i^0 \tilde{X}_i^0(\mathbf{x}_j^0) \tilde{T}_i^0(t_j^0), \quad j = 1, \ldots, N^0,$$

that represents a linear system of size N^0 whose solution allows calculating the coefficients ξ_i^0.

With the coefficients ξ_i^0 known, the nonlinear term is approximated by

$$\mathcal{L}^0 = \sum_{i=1}^{i=N^0} \xi_i^0 \tilde{X}_i^0(\mathbf{x}) \tilde{T}_i^0(t).$$

Now, the next iteration solves

$$\mathcal{L}^L(u^1) = \mathcal{L}^0 + f(\mathbf{x}, t),$$

with $u^1(\mathbf{x}, t)$ recomputed from scratch or by adding new terms to $u^0(\mathbf{x}, t)$.

The new separated representation of $u^1(\mathbf{x}, t)$, involving N^1 terms, allows calculating the new basis $\mathcal{B}^1 = \{\tilde{X}_1^1 \tilde{T}_1^1, \ldots, \tilde{X}_{N^1}^1 \tilde{T}_{N^1}^1\}$ and from it, approximating the nonlinear term by interpolating at the N^1 points $(\mathbf{x}_j^1, t_j^1)$, $j = 1, \ldots, N^1$.

The only point that deserves additional comments is the one related to the choice of the interpolation points $(\mathbf{x}_j^k, t_j^k)$, $j = 1, \ldots, N^k$ at iteration k.

At the kth-iteration the reduced approximation basis reads $\mathcal{B}^k = \{\tilde{X}_1^k \tilde{T}_1^k, \ldots, \tilde{X}_{N^k}^k \tilde{T}_{N^k}^k\}$. Thus, we consider

$$(\mathbf{x}_1^k, t_1^k) = \arg\max_{\mathbf{x}, t} |\tilde{X}_1^k(\mathbf{x}) \tilde{T}_1^k(t)|,$$

then we compute d_1 to ensure

$$d_1 \tilde{X}_1^k(\mathbf{x}_1^k) \tilde{T}_1^k(t_1^k) = \tilde{X}_2^k(\mathbf{x}_1^k) \tilde{T}_2^k(t_1^k),$$

that allows defining $r_2^k(\mathbf{x}, t)$

$$r_2^k(\mathbf{x}, t) = \tilde{X}_2^k(\mathbf{x}) \tilde{T}_2^k(t) - d_1 \tilde{X}_1^k(\mathbf{x}) \tilde{T}_1^k(t),$$

and from it computing point $(\mathbf{x}_2^k, t_2^k)$ according to

$$(\mathbf{x}_2^k, t_2^k) = \arg\max_{\mathbf{x}, t} |r_2^k(\mathbf{x}, t)|.$$

Since by construction $r_2^k(\mathbf{x}_1^k, t_1^k) = 0$, we can ensure $(\mathbf{x}_2^k, t_2^k) \neq (\mathbf{x}_1^k, t_1^k)$.

The procedure is generalized to obtain the other points involved in the interpolation.

References

1. A. Ammar, B. Mokdad, F. Chinesta, R. Keunings, A new family of solvers for some classes of multidimensional partial differential equations encountered in kinetic theory modeling of complex fluids. J. Non-Newtonian Fluid Mech. **139**, 153–176 (2006)
2. A. Ammar, E. Cueto, F. Chinesta, Reduction of the chemical master equation for gene regulatory networks using proper generalized decompositions. Int. J. Numer. Methods Biomed. Eng. **28**(9), 960–973 (2012)

3. P. Ladeveze, The large time increment method for the analyze of structures with nonlinear constitutive relation described by internal variables. CRAS **309**, 1095–1099 (1989)
4. P. Ladeveze, *Nonlinear Computational Structural Mechanics* (Springer, New Approaches and Non-Incremental Methods of Calculation, 1999)
5. A. Ammar, B. Mokdad, F. Chinesta, R. Keunings, Transient simulation using space-time separated representation. A new family of solvers for some classes of multidimensional partial differential equations encountered in kinetic theory modeling of complex fluids. Part II. J. Non-Newtonian Fluid Mech. **144**, 98–121 (2007)
6. G. Bonithon, P. Joyot, F. Chinesta, P. Villon, Non-incremental boundary element discretization of parabolic models based on the use of proper generalized decompositions. Eng. Anal. Boundary Elements **35**(1), 2–17 (2011)
7. B. Bognet, A. Leygue, F. Chinesta, A. Poitou, F. Bordeu, Advanced simulation of models defined in plate geometries: 3D solutions with 2D computational complexity. Comput. Methods Appl. Mech. Eng. **201**, 1–12 (2012)
8. B. Bognet, A. Leygue, F. Chinesta, Separated representations of 3D elastic solutions in shell geometries. Adv. Model. Simul. Eng. Sci. **1**, 4 (2014)
9. A. Ammar, F. Chinesta, A. Falco, On the convergence of a greedy rank-one update algorithm for a class of linear systems. Arch. Comput. Methods Eng. **17**(4), 473–486 (2010)
10. D. Gonzalez, E. Cueto, F. Chinesta, Real-time direct integration of reduced solid dynamics equations. Int. J. Numer. Methods Eng. **99**(9), 633–653 (2014)
11. A. Ammar, F. Chinesta, P. Diez, A. Huerta, An error estimator for separated representations of highly multidimensional models. Comput. Methods Appl. Mech. Eng. **199**, 1872–1880 (2010)
12. E. Nadal, A. Leygue, F. Chinesta, M. Beringhier, J.J. Rodenas, F.J. Fuenmayor, A separated representation of an error indicator for the mesh refinement process under the Proper Generalized Decomposition framework. Comput. Mech. **55**(2), 251–266 (2015)
13. I. Alfaro, D. Gonzalez, S. Zlotnik, P. Diez, E. Cueto, F. Chinesta, An error estimator for real-time simulators based on model order reduction. Adv. Model. Simul. Eng. Sci. **2**, 30 (2015)
14. F. Chinesta, A. Leygue, F. Bordeu, J.V. Aguado, E. Cueto, D. Gonzalez, I. Alfaro, A. Ammar, A. Huerta, Parametric PGD based computational vademecum for efficient design, optimization and control. Archives Comput. Methods Eng. **20**(1), 31–59 (2013)
15. A. Ammar, A. Huerta, F. Chinesta, E. Cueto, A. Leygue, Parametric solutions involving geometry: a step towards efficient shape optimization. Comput. Methods Appl. Mech. Eng. **268C**, 178–193 (2014)
16. M. Nazeer, F. Bordeu, A. Leygue, F. Chinesta, Arlequin based PGD domain decomposition. Comput. Mech. **54**(5), 1175–1190 (2014)
17. A. Huerta, E. Nadal, F. Chinesta, Domain decomposition and the proper generalized decomposition. Int. J. Numer. Methods Eng. **113**(13), 1972–1994 (2018)
18. A. Ammar, F. Chinesta, E. Cueto, Coupling finite elements and proper generalized decompositions. Int. . Multiscale Comput. Eng. **9**(1), 17–33 (2011)
19. D. Gonzalez, E. Cueto, F. Chinesta, P. Diez, A. Huerta, SUPG-based stabilization of proper generalized decompositions for high-dimensional advection-diffusion equations. Inte. J. Numer. Methods Eng. **94**(13), 1216–1232 (2013)
20. A. Leygue, F. Chinesta, M. Beringhier, T.L. Nguyen, J.C. Grandidier, F. Pasavento, B. Schrefler, Towards a framework for non-linear thermal models in shell domains. Int. J. Numer. Methods Heat Fluid Flow **23**(1), 55–73 (2013)
21. J.V. Aguado, D. Borzacchiello, K.S. Kollepara, F. Chinesta, A. Huerta, Tensor representation of on-linear models using cross approximations. J. Scientific Comput. **81**, 22–47 (2019)
22. M. Beringhier, A. Leygue, F. Chinesta, Parametric nonlinear PDEs with multiple solutions: a PGD approach. Discrete Continuous Dyn. Syst. **9**(2), 383–392 (2016)
23. F. Chinesta, P. Ladeveze, E. Cueto, A short review in model order reduction based on Proper Generalized Decomposition. Archives Comput. Methods Eng. **18**, 395–404 (2011)

24. F. Chinesta, R. Keunings, A. Leygue, *The Proper Generalized Decomposition for Advanced Numerical Simulations* (Springer, A primer. Springerbriefs, 2014)
25. F. Chinesta, A. Huerta, G. Rozza, K. Willcox, Model order reduction, in *the Encyclopedia of Computational Mechanics, Second Edition, Erwin Stein, Rene de Borst, Tom Hughes Edt.*, John Wiley & Sons, Ltd. (2015)

Open Access This chapter is licensed under the terms of the Creative Commons Attribution-NonCommercial-NoDerivatives 4.0 International License (http://creativecommons.org/licenses/by-nc-nd/4.0/), which permits any noncommercial use, sharing, distribution and reproduction in any medium or format, as long as you give appropriate credit to the original author(s) and the source, provide a link to the Creative Commons license and indicate if you modified the licensed material. You do not have permission under this license to share adapted material derived from this chapter or parts of it.

The images or other third party material in this chapter are included in the chapter's Creative Commons license, unless indicated otherwise in a credit line to the material. If material is not included in the chapter's Creative Commons license and your intended use is not permitted by statutory regulation or exceeds the permitted use, you will need to obtain permission directly from the copyright holder.

Chapter 25
Design of Experiments and Surrogate Models

25.1 Design of Experiments, DoE

25.1.1 The Multifactorial Approach of Plackett and Burman and Factors Influence

DoE is much larger than its application in the construction of surrogate models. DoEs were first used in reliability analyses, for example to ascertain the effect of quantitative or qualitative alterations in the various components upon some measured characteristic of the complete assembly.

In the design of optimum multifactorial experiments by Plackett and Burman [1], it is considered that the output y depends on n factors x_i, $i = 1, \ldots, n$.

When considering two levels, factor x_i has only two possible values (in the original work the authors considered more levels). The so-called nominal configuration is given by the factors choice $(x_1^0, \ldots, x_n^0)$.

Thus, the influence of factor x_i could be expressed from the quantity m_i making use of the altered value of the i-factor, x_i',

$$m_i = \frac{\sum y(x_1, \ldots, x_{i-1}, x_i', x_{i+1}, \ldots, x_n) - \sum y(x_1, \ldots, x_{i-1}, x_i^0, x_{i+1}, \ldots, x_n)}{2^n},$$

where 2^n is the number of all the possible cases, and the sums applies for all the combinations of all the factors except x_i.

The cross influence between factors x_i and x_j, m_{ij}, can be defined in a similar way. The grand average M is the average of the 2^n combinations.

25.1.2 Fisher Information Matrix, FIM

The probability distribution function of X with respect to parameter μ, reads $f(X;\mu)$. The main issue in defining a DoE, is the ease to identify parameters by maximizing likelihood. When the likelihood is flat, many samples are needed to determine the actual parameters value with an adequate accuracy. In that case, the variance with respect to μ could be a valuable indicator to optimize the design.

The so-called *score* is defined from $\frac{\partial}{\partial \mu} \log f(X;\mu)$. Its first moment vanishes under certain regularity conditions. In fact

$$\mathbb{E}\left(\frac{\partial}{\partial \mu} \log f(X;\mu)\right) = \int \frac{\frac{\partial}{\partial \mu} f(x;\mu)}{f(x;\mu)} f(x;\mu) dx = \frac{\partial}{\partial \mu} \int f(x;\mu) dx = \frac{\partial}{\partial \mu} 1 = 0,$$

and the variance of the score is by definition the Fisher information $\mathcal{I}(\mu)$

$$\mathcal{I}(\mu) = \mathbb{E}\left(\left(\frac{\partial}{\partial \mu} \log f(X;\mu)\right)^2\right),$$

which, taking into account

$$\frac{\partial^2}{\partial \mu^2} \log f(X;\mu) = \frac{\frac{\partial^2 f(X;\mu)}{\partial \mu^2}}{f(X;\mu)} - \left(\frac{\partial}{\partial \mu} \log f(X;\mu)\right)^2,$$

and

$$\mathbb{E}\left(\frac{\frac{\partial^2 f(X;\mu)}{\partial \mu^2}}{f(X;\mu)}\right) = \frac{\partial^2}{\partial \mu^2} \int f(x;\mu) dx = 0,$$

yields

$$\mathcal{I}(\mu) = -\mathbb{E}\left(\frac{\partial^2}{\partial \mu^2} \log f(X;\mu)\right).$$

When considering many parameters $\mu_1, \mu_2, \ldots$, it results in the so-called Fisher information matrix, whose components read

$$\mathcal{I}_{i,j}(\boldsymbol{\mu}) = \mathbb{E}\left(\left(\frac{\partial}{\partial \mu_i} \log f(X, \boldsymbol{\mu})\right)\left(\frac{\partial}{\partial \mu_j} \log f(X, \boldsymbol{\mu})\right)\right),$$

which under usual regularity conditions can be rewritten as

$$\mathcal{I}_{i,j}(\boldsymbol{\mu}) = -\mathbb{E}\left(\frac{\partial^2}{\partial \mu_i \partial \mu_j} \log f(X, \boldsymbol{\mu})\right).$$

Fisher information based active learning sample the parametric space driven by the Fisher matrix.

25.1.3 Quadratures and Latin-Hypercube

Some of the techniques considered in the next sections make use of DoE based on quadratures. The quadrature points are the roots of a polynomial belonging to a class of orthogonal polynomials (orthogonal with respect to a weighted inner-product). In the multidimensional case, the tensor product of the one-dimensional bases applies.

If we consider for the sake of simplicity a 2D case, the three simplest samplings consist of

- Random sampling, that adds points without taking into account the ones already placed in the parametric domain;
- Latin Hypercube divide the domain in a series of rows and columns, and when considering a tentative point related to a random row and a column, it is accepted if the row and the columns do not contain any point;
- A variant of the Latin Hypercube is the so-called orthogonal sampling, in which the domain is partitioned in equally probable regions, and a point is accepted if it defines a Latin Hypercube sampling and moreover, each region is sampled with the same density of points.

Surrogate models usually consider Latin Hypercube based samplings, even if nowadays, active learning looks for more valuable samplings, as just commented in the previous section. Gaussian processes are at the origin of different active learning procedures.

25.2 Surrogate Models Based on Response Surface Methodology, RSM

Response surface is a traditional methodology for, from some calculations of a given parametric problem for given choices of the model parameters $\mathbf{p}_i \in \mathbb{R}^P$, $i = 1, \ldots, M$, $\mathbf{u}_i \equiv \mathbf{u}(\mathbf{p}_i)$, compute the solution for any other value of the parameter $\mathbf{p}$ [2, 3]. RSM make use of approximations or interpolations

$$\mathbf{u}(\mathbf{p}) = \sum_{i=1}^{M} \mathbf{u}_i \mathcal{F}_i(\mathbf{p}),$$

with $\mathcal{F}_i(\mathbf{p})$ and adequate approximation (interpolation as soon as $\mathcal{F}_i(\mathbf{p}_j) = \delta_{ij}$, with δ_{ij} the Kronecker delta).

Such a parametric function $\mathbf{u}(\mathbf{p})$ defines a manifold, also known as response surface.

The choice of $\mathcal{F}_i(\mathbf{p})$ is the most delicate task. It should enable to operate with a sparse distribution of data, produce smooth enough approximations, be robust with respect to outliers and avoid overfitting as much as possible.

Orthogonal polynomials offer excellent behaviors however they impose the necessity of associating the sampling to the so-called Gauss-Lobatto quadratures. This is not easy to accomplish in general, when the parametric domain has a complex shape, making difficult a tensor product of the approximation bases and the cartesian product of their associated collocation points.

When the points are distributed randomly, usual polynomial bases perform very poorly, with large spurious oscillations. In those cases low degree polynomials are preferred because their use limits these effects, with the price of losing the interpolation behavior.

A very appealing technique able to proceed smoothly, in general sparse data-sets, with nonuniform distributions, is the so-called *kriging*, revisited in Chaps. 14 and 19. However, usual RSM techniques, rapidly fail when the parametric dimensionality increases.

References

1. R.L. Plackett, J.P. Burman, The design of optimum multifactorial experiments. Biometrika **33**, 305–325 (1946)
2. P. Jiang, Q. Zhou, X. Shao, *Surrogate Model-Based Engineering Design and Optimization* (Springer, 2020)
3. A.I.J. Forrester, A. Sobester, A.J. Keane, *Engineering Design via Surrogate Modelling: A Practical Guide* (John Wiley & Sons, Ltd, 2008)

Open Access This chapter is licensed under the terms of the Creative Commons Attribution-NonCommercial-NoDerivatives 4.0 International License (http://creativecommons.org/licenses/by-nc-nd/4.0/), which permits any noncommercial use, sharing, distribution and reproduction in any medium or format, as long as you give appropriate credit to the original author(s) and the source, provide a link to the Creative Commons license and indicate if you modified the licensed material. You do not have permission under this license to share adapted material derived from this chapter or parts of it.

The images or other third party material in this chapter are included in the chapter's Creative Commons license, unless indicated otherwise in a credit line to the material. If material is not included in the chapter's Creative Commons license and your intended use is not permitted by statutory regulation or exceeds the permitted use, you will need to obtain permission directly from the copyright holder.

Chapter 26
Parametric Models: POD Based Surrogates

Even when reduced-order bases can be determined for a given problem, the just seen procedures work by projection of the problem onto the subspace spanned by these bases (standard POD formulation, already described in Sect. 23). This makes it difficult to obtain results under severe real-time feedback constraints. In that case surrogates are often preferred.

26.1 POD with Interpolation

The so-called POD with interpolation, PODI [1], lies at the origin of the non-intrusive POD. PODI employs snapshots related to different values of the model parameter p, $\mathbf{u}(p_i)$, $i = 1, \ldots, M$. We assume here that this parameter is a scalar (the multiparametric case will be addressed later on) and that they are ordered, i.e., $p_1 < \cdots < p_M$.

As we did in standard POD-based MOR, we extract the reduced basis, $\phi_1, \ldots, \phi_R$. Given parameter p, with $p \neq \{p_1, p_2, \ldots, p_M\}$, instead of projecting the searched solution onto the reduced basis $\mathbf{u}(p) = \sum_{i=1}^{R} \gamma_i(p) \phi_i$, and then calculating the coefficient $\gamma_i(p)$ by Galerkin projection, i.e., by solving $(\mathbf{B}^T \mathbf{K} \mathbf{B}) \gamma = \mathbf{B}^T \mathbf{F}$ (that needs the matrix assembling, the matrix products before finally solving the reduced linear system of equations), PODI proceeds by

- *Sampling*: $\mathbf{u}(p_i) \equiv \mathbf{u}_i$, $i = 1, \ldots, M$;
- *Reduced basis extraction*: POD is applied to extract the reduced basis $\phi_1, \ldots, \phi_R$;
- *Reproduction*: calculation of γ_i. For that, one needs to express $\mathbf{u}_i = \sum_{j=1}^{R} \gamma_j^i \phi_j$. Premultiplying by ϕ_k and taking into account the orthonormality of the reduced basis, we obtain

$$\phi_k^T \mathbf{u}_i = \gamma_k^i.$$

Repeating for all $i \in \{1, \ldots, M\}$ and $k \in \{1, \ldots, R\}$, we obtain γ_i (the reduced counterpart of $\mathbf{u}_i$).

- *Interpolation*: With the reduced solution representations $\gamma_i \equiv \gamma(p = p_i)$, one is tempted for any other p to proceed by approximation or interpolation, i.e.

$$\gamma(p) = \sum_{i=1}^{R} \gamma_i \mathcal{F}_i(p),$$

with $\mathcal{F}_i(p)$ the approximation functions, that define an interpolation as soon as $\mathcal{F}_i(p_j) = \delta_{ij}$, with δ_{ij} the Kronecker delta.

- *Reconstruction*: With the $\gamma(p)$ just obtained, the solution can be reconstructed everywhere by iusing $\mathbf{u}(p) = \mathbf{B}\gamma(p)$.

26.2 Extension to Multi-parametric Scenarios

The extension of the just presented technique to highly-multidimensional frameworks becomes difficult because usual approximation bases suffer from the so-called curse of dimensionality.

In the case of moderate dimensionality, however, the PODI algorithm is easily generalizable. To this end, we first reformulate the PODI described above as follows: the reconstruction $\mathbf{u}(p) = \mathbf{B}\gamma(p)$ can be expressed in the equivalent form:

$$\mathbf{u}(p) = \sum_{k=1}^{R} \gamma_k(p) \phi_k;$$

with $\gamma_k^i \equiv \gamma_k(p_i)$ known, the interpolation can be expressed as:

$$\mathbf{u}(p) = \sum_{k=1}^{R} \left(\sum_{i=1}^{M} \gamma_k^i \mathcal{F}_i(p) \right) \phi_k.$$

This is generalizable to the multi-parametric case, where the scalar p is replaced by a parameters vector $\mathbf{p}$, with the interpolation expressed now as

$$\mathbf{u}(\mathbf{p}) = \sum_{k=1}^{R} \left(\sum_{i=1}^{M} \gamma_k^i \mathcal{F}_i(\mathbf{p}) \right) \phi_k.$$

When the number of parameters (the size of vector $\mathbf{p}$) grows up, the interpolation process becomes costly. Separated representations in sparse settings, addressed in Sect. 27, succeed in circumventing the just referred difficulty.

Reference

1. H.V. Ly, H.T. Tran, Modeling and control of physical processes using Proper Orthogonal Decomposition. J. Math. Comput. Model. **33**(1–3), 223–236 (2001)

Open Access This chapter is licensed under the terms of the Creative Commons Attribution-NonCommercial-NoDerivatives 4.0 International License (http://creativecommons.org/licenses/by-nc-nd/4.0/), which permits any noncommercial use, sharing, distribution and reproduction in any medium or format, as long as you give appropriate credit to the original author(s) and the source, provide a link to the Creative Commons license and indicate if you modified the licensed material. You do not have permission under this license to share adapted material derived from this chapter or parts of it.

The images or other third party material in this chapter are included in the chapter's Creative Commons license, unless indicated otherwise in a credit line to the material. If material is not included in the chapter's Creative Commons license and your intended use is not permitted by statutory regulation or exceeds the permitted use, you will need to obtain permission directly from the copyright holder.

Chapter 27
Parametric Models: PGD-Based Surrogates

The present section revisits some technologies for constructing surrogates, all them based on the use of separated representations, at the heart of the so-called PGD.

27.1 Sparse Subspace Learning

We consider the general case of a time-dependent parametric problem. We assume, again for simplicity, that only one parameter is involved in the model, $\mu \in \Omega_\mu = [\mu_{\min}, \mu_{\max}]$. Generalizing to several parameters is conceptually straightforward, but we must keep in mind that, given the existence of the so-called curse of dimensionality, ubiquitous in this book, this setting cand not be ignored. The sought parametric solution $u(\mathbf{x}, t, \mu)$ is to be constructed in the separated form

$$u(\mathbf{x}, t, \mu) \approx \sum_{i=1}^{N} X_i(\mathbf{x}, t) M_i(\mu). \tag{27.1}$$

Space-time and parameter functions, $X_i(\mathbf{x}, t)$ and $M_i(\mu)$ respectively, are a priori unknown. Sparse Subspace Learning (SSL) techniques [1, 2] first choose a hierarchical basis of the parametric domain. The collocation points (using a Gauss-Lobatto quadrature) and their associated functions will be denoted by: $(\mu_i^j, \xi_i^j(\mu))$, where indexes i and j refer to the ith point at the jth level.

When using Chebyshev polynomials, at the level $j = 0$ there are only two points, μ_1^0 and μ_2^0. They correspond to the minimum and maximum value of the parameter that define the parametric domain, i.e. $\mu_1^0 = \mu_{\min}$ and $\mu_2^0 = \mu_{\max}$.

If we assume that a direct solver is available, i.e., a computer software able to find the transient solution once the value of the parameter has been specified, solutions associated to both collocation points μ_1^0 and μ_2^0 expressed as $u_1^0(\mathbf{x}, t) = u(\mathbf{x}, t, \mu = \mu_1^0)$ and $u_2^0(\mathbf{x}, t) = u(\mathbf{x}, t, \mu = \mu_2^0)$, respectively.

Thus, the solution at level $j = 0$ could be approximated as

$$u^0(\mathbf{x}, t, \mu) = u_1^0(\mathbf{x}, t)\xi_1^0(\mu) + u_2^0(\mathbf{x}, t)\xi_2^0(\mu),$$

that has the separated structure given by Eq. (27.1). In fact, the solution at level $j = 0$ consists of a standard linear approximation since at the first level, the two approximation functions read $\xi_1^0(\mu) = (\mu_2^0 - \mu)/(\mu_2^0 - \mu_1^0)$ and $\xi_2^0(\mu) = (\mu - \mu_1^0)/(\mu_2^0 - \mu_1^0)$, respectively.

At level $j = 1$ only one point exists and is located at the midpoint of the parametric domain, i.e. $\mu_1^1 = 0.5(\mu_{\min} + \mu_{\max})$. Its associated interpolation function is $\xi_1^1(\mu)$. $\xi_1^1(\mu)$ is a parabola that takes a unit value at $\mu = \mu_1^1$ and vanishes at the other collocation points of level $j = 0$, μ_1^0 and μ_2^0. The associated solution reads

$$u_1^1(\mathbf{x}, t) = u(\mathbf{x}, t, \mu = \mu_1^1).$$

This solution contains a part already explained by the just computed approximation at the previous level, $j = 0$, expressed by $u^0(\mathbf{x}, t, \mu_1^1) = u_1^0(\mathbf{x}, t)\xi_1^0(\mu_1^1) + u_2^0(\mathbf{x}, t)\xi_2^0(\mu_1^1)$.

The so-called *surplus* $\tilde{u}_1^1(\mathbf{x}, t)$ is defined as

$$\tilde{u}_1^1(\mathbf{x}, t) = u_1^1(\mathbf{x}, t) - u^0(\mathbf{x}, t, \mu_1^1),$$

from which, the approximation at level $j = 1$ reads

$$u^1(\mathbf{x}, t, \mu) = u^0(\mathbf{x}, t, \mu) + \tilde{u}_1^1(\mathbf{x}, t)\xi_1^1(\mu).$$

We continue adding surpluses as the hierarchical approximation level increases. It is worth highlighting that the norm of the surplus can be used as a local error indicator.

Consider the two-parameter case, i.e., when μ and η are two parameters governing the solution. The hierarchical two-dimensional basis in the parameter space (μ, η) is composed by the cartesian product of the collocations points and the tensor product of the approximation bases. Thus, the first level $j = 0$, is now composed by four points:

$$(\mu_1^0, \eta_1^0), \ (\mu_2^0, \eta_1^0), \ (\mu_2^0, \eta_2^0) \text{ and } (\mu_1^0, \eta_2^0),$$

whose corresponding interpolation functions are:

$$\xi_1^0(\mu)\varphi_1^0(\varphi), \ \xi_2^0(\mu)\varphi_1^0(\eta), \ \xi_2^0(\mu)\varphi_2^0(\eta) \text{ and } \xi_1^0(\mu)\varphi_2^0(\eta).$$

Level-one solution combines level-zero solution of one of the coordinates with the one in the other, and so on. As already introduced before, this algorithm becomes too expensive when either the number of parameters or the required degree increase.

If we consider P parameters, the first approximation level, $j = 0$, already involves 2^P collocation points.

Another weakness is associated with the fact that the Gauss-Lobatto quadrature points associated with Chebyshev polynomials, are densely distributed in the parametric domain boundary neighborhood, quite far from the center that in general represents the nominal parameters. Thus, when considering only the first approximation levels, the accuracy around the nominal conditions is quite poor.

27.2 Collocation-Based Sparse PGD and Its Regularized Variants

Consider again a two-parameter problem whose solution is $u(\mathbf{x}, t, \mu, \eta)$. We assume that we know in advance n_t solutions related to n_t different choices of the parameters. In other words, $u(\mathbf{x}, t, \mu_i, \eta_i) \equiv u^i(\mathbf{x}, t), i = 1, \ldots, n_t$.

These known solutions are stored in the form of a matrix, whose columns contain the solution evaluated at each node (rows) at a given time step. The discretized, matrix form of the solution $u(\mathbf{x}, t, \mu_i, \eta_i) \equiv u^i(\mathbf{x}, t)$ will hereafter be denoted by $\mathbb{U}^i$, with the component $\mathbb{U}^i_{rs}$ referring to the solution at node $\mathbf{x}_r$ at time t_s, for the parameter values (μ_i, η_i), i.e. $\mathbb{U}^i_{rs}$ represents $u(\mathbf{x}_r, t_s, \mu_i, \eta_i)$.

Finding a parametric regression for each component of the discrete solution leads to $\mathbb{U}(\mu, \eta)$. This, however, becomes easily too expensive and losses smoothness. For this reason, instead of working with the discrete solution itself, reduced space and time bases are extracted: $\mathbf{X}_k, k = 1, \ldots, K$, and $\mathbf{T}_l, l = 1, \ldots, L$. In this way, each solution vector $\mathbb{U}^i$ can be expressed as

$$\mathbb{U}^i \approx \mathbb{X}\mathbb{D}^i\mathbb{T},$$

where $\mathbb{X} = (\mathbf{X}_1 \ldots \mathbf{X}_K)$, $\mathbb{T}^T = (\mathbf{T}_1 \ldots \mathbf{T}_L)$ and $\mathbb{D}$, with size $K \times L$, is obtained by minimizing $\|\mathbb{U}^i - \mathbb{X}\mathbb{D}^i\mathbb{T}\|$.

Assume now that $\mathbb{D}^i_{pq}$ is known. In other words, we know each component for each sampling $i, i = 1, \ldots, n_t$, of the parameters, (μ_i, η_i). Therefore, $\mathbb{D}^i_{pq} \equiv \mathbb{D}_{pq}(\mu_i, \eta_i)$, a general parametric expression $\mathbb{D}_{pq}(\mu, \eta)$ is searched.

Without loss of generality, we describe the approximation of a scalar function f that depends on the pair of parameters (μ, η). We assume $f^i \equiv f(\mu_i, \eta_i)$ are known at the n_t points of the sampling campaign. Here, the scalar f represents each component $\mathbb{D}_{pq}, p = 1, \ldots, K$ and $q = 1, \ldots, L$. The goal is therefore to find a function $f^M(\mu, \eta)$, expressible as a finite sum of M terms, according to

$$f^M(\mu, \eta) = \sum_{j=1}^{M} \mathcal{G}_j(\mu)\mathcal{H}_j(\eta),$$

able to approximate the known data $f^i \equiv f(\mu_i, \eta_i), i = 1, \ldots, n_t$.

Note that in two-dimensional scenarios we do not need, in general, to construct separated representations. However, our aim here is to introduce the technique for a general enough problem, eventually involving numerous parameters.

The proposed greedy algorithm works by looking, at iteration m, for the update $\mathcal{G}_m(\mu)\mathcal{H}_m(\eta)$ such that $f^m(\mu,\eta) = f^{m-1}(\mu,\eta) + \mathcal{G}_m(\mu)\mathcal{H}_m(\eta)$. To compute the sought functions, we first approximate them as $\mathcal{G}_m(\mu) = \mathbf{N}_m^{\mu,T}(\mu)\mathbf{a}_m$ and $\mathcal{H}_m(\eta) = \mathbf{N}_m^{\eta,T}(\eta)\mathbf{b}_m$, where $\mathbf{N}_m^{\mu}$ represents the basis considered for approximating the m-th mode, that depends on the μ-parameter, with $\mathbf{a}_m$ the associated weights. We proceed similarly for the other direction (η-parameter): $\mathbf{N}_m^{\eta}$ and $\mathbf{b}_m$.

The update is found by solving the minimization problem:

$$\mathcal{G}_m(\mu)\mathcal{H}_m(\eta) = \arg\min_{(\mathcal{G}_m(\mu)\mathcal{H}_m(\eta))^*} \sum_{i=1}^{n_t} \|f^i - f^{m-1} + (\mathcal{G}_m(\mu)\mathcal{H}_m(\eta))^*\|_2^2,$$

that, by using the notation below, reads

$$\mathbf{r} = \begin{pmatrix} f^1 - f^{m-1}(\mu_1,\eta_1) \\ \vdots \\ f^{n_t} - f^{m-1}(\mu_{n_t},\eta_{n_t}) \end{pmatrix},$$

$$\mathbb{M}_\mu = \begin{pmatrix} \mathbf{N}_m^{\eta,T}(\eta_1)\mathbf{b}_m \mathbf{N}_m^{\mu,T}(\mu_1)) \\ \vdots \\ \mathbf{N}_m^{\eta,T}(\eta_{n_t})\mathbf{b}_m \mathbf{N}_m^{\mu,T}(\mu_{n_t})) \end{pmatrix},$$

$$\mathbb{M}_\eta = \begin{pmatrix} \mathbf{N}_m^{\mu,T}(\mu_1)\mathbf{a}_m \mathbf{N}_m^{\eta,T}(\eta_1)) \\ \vdots \\ \mathbf{N}_m^{\mu,T}(\mu_{n_t})\mathbf{a}_m \mathbf{N}_m^{\eta,T}(\eta_{n_t})) \end{pmatrix},$$

results in the two problems:

$$\mathbf{a}_m = \arg\min_{\mathbf{a}_m^*} \left\{ \|\mathbf{r} - \mathbb{M}_\mu \mathbf{a}_m^*\|_2^2 \right\}, \tag{27.2}$$

$$\mathbf{b}_m = \arg\min_{\mathbf{b}_m^*} \left\{ \|\mathbf{r} - \mathbb{M}_\eta \mathbf{b}_m^*\|_2^2 \right\}. \tag{27.3}$$

This is solved iteratively until a fixed point, whose existence is assumed, is found.

This algorithm is at the heart of the sparse-PGD (sPGD) constructor [3]. However, overfitting appears when combining rich approximation bases $\mathbf{N}_m^{\mu}(\mu)$ and $\mathbf{N}_m^{\eta}(\eta)$, with data-sets n_t not rich enough. This will be always the case when operating in highly multi-parametric settings.

To avoid overfitting, an adaptive procedure was proposed in [3] that consists in adapting the approximation bases, such that their degree increases when advancing in the modal enrichment m. Looking for a more versatile and automatic method, different regularizations were also proposed in [4].

27.2.1 Regularized Formulations

The so-called *ridge* regularization reads

$$\mathbf{a}_m = \arg\min_{\mathbf{a}_m^*} \left\{ \|\mathbf{r} - \mathbb{M}_\mu \mathbf{a}_m^*\|_2^2 + \lambda \|\mathbf{a}_m^*\|_2^2 \right\}, \tag{27.4}$$

$$\mathbf{b}_m = \arg\min_{\mathbf{b}_m^*} \left\{ \|\mathbf{r} - \mathbb{M}_\eta \mathbf{b}_m^*\|_2^2 + \lambda \|\mathbf{b}_m^*\|_2^2 \right\}, \tag{27.5}$$

where the L2 regularization attempts to prevent overfitting.

When looking for sparsity in the approximation, in order to employ extremely rich approximations while maintaining parsimony (something already proposed in the SINDy method [5]), it is well known that the lower is the employed norm, the more intense becomes the enforcement of sparsity. Very often the L1-norm (the so-called *Lasso* regularization) becomes an appealing choice, a compromise between sparsity enforcement and computational efficiency. In turn, if we combine *ridge* and *Lasso*, giving rise to the so-called *elastic net* regularization, this results in the separated representation setting [4]

$$\mathbf{a}_m = \arg\min_{\mathbf{a}_m^*} \left\{ \|\mathbf{r} - \mathbb{M}_\mu \mathbf{a}_m^*\|_2^2 + \lambda \left[(1-\alpha)\|\mathbf{a}_m^*\|_2^2 + \alpha \|\mathbf{a}_m^*\|_1\right] \right\}, \tag{27.6}$$

$$\mathbf{b}_m = \arg\min_{\mathbf{b}_m^*} \left\{ \|\mathbf{r} - \mathbb{M}_\eta \mathbf{b}_m^*\|_2^2 + \lambda \left[(1-\alpha)\|\mathbf{b}_m^*\|_2^2 + \alpha \|\mathbf{b}_m^*\|_1\right] \right\}. \tag{27.7}$$

Figure 27.1 summarizes the different possibilities at hand for regression techniques. They conciliate: (i) a sparse and very reduced sampling; (ii) rich enough approximation bases while avoiding overfitting, based on the use of sparse regularizations; (iii) using orthogonal basis for evaluating sensibilities in a direct manner; and (iv) efficiently addressing the high-dimensional spaces induced by the multi-parametric models, where the use of separated representations are specially suitable [4].

27.2.2 Smaller Than the L1 Norm

The L1-norm is usually considered for enforcing sparsity, of particular interest in sparse identification for constructing regressions allying richness and parsimony, to prevent overfitting.

In sparse identification, the L1-norm is considered because of the computational difficulties related to the use of the most suitable, from the sparsity viewpoint, L0-norm. However, in between the L0 and the L1, the so-called Lp-norm can be a nice compromise, enhancing sparsity while alleviating the computational difficulties.

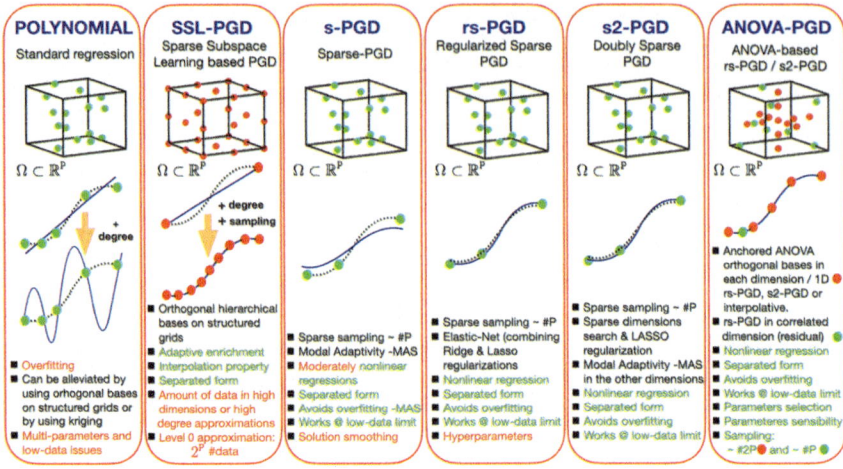

Fig. 27.1 Advanced nonlinear regressors

The Lp minimization problem can be formulated as follows:

$$\begin{cases} \min_{\mathbf{s}} \tilde{F}(\mathbf{s}) = \min_{\mathbf{s}} \sum_{i=1}^{n} |s_i|^p \\ \text{subjected to: } \mathbf{As} = \mathbf{x} \end{cases}.$$

The so-called FOCUSS algorithm proceeds by iterating according to:

$$\mathbf{s}^{m+1} = \mathbf{P}^{-1}(\mathbf{s}^m)\mathbf{A}^T\{\mathbf{A}\mathbf{P}^{-1}(\mathbf{s}^m)\mathbf{A}^T\}^{-1}\mathbf{x}, \quad m = 0, 1, \ldots,$$

with

$$\mathbf{P}(\mathbf{s}^m) = \mathtt{diag}(|s_1^m|^{p-2}, \ldots, |s_n^m|^{p-2}).$$

The non rigorous (in the sense discussed in [6]) derivation of the FOCUSS algorithm uses a Lagrange multiplier for defining the Lagrangian $L(\mathbf{s}, \boldsymbol{\alpha})$: $L(\mathbf{s}, \boldsymbol{\alpha}) = \tilde{F}(\mathbf{s}) + \boldsymbol{\alpha}^T(\mathbf{As} - \mathbf{x})$, whose extrema read:

$$\begin{cases} \frac{\partial L(\mathbf{s}, \boldsymbol{\alpha})}{\partial \mathbf{s}} = \frac{\partial \tilde{F}(\mathbf{s})}{\partial \mathbf{s}} + \mathbf{A}^T \boldsymbol{\alpha} = \mathbf{0} \\ \frac{\partial L(\mathbf{s}, \boldsymbol{\alpha})}{\partial \boldsymbol{\alpha}} = \mathbf{As} - \mathbf{x} = \mathbf{0} \end{cases}, \qquad (27.8)$$

where $\frac{\partial \tilde{F}(\mathbf{s})}{\partial \mathbf{s}} = p\,\mathbf{P}(\mathbf{s})\,\mathbf{s}$, that leads to the FOCUSS linearized iteration algorithm. For that, it suffices to proceed as follows:

- The first expression in Eq. (27.8) leads to: $\mathbf{s} = -\frac{1}{p}\mathbf{P}^{-1}(\mathbf{s})\mathbf{A}^T\boldsymbol{\alpha}$;
- Then, using the second one with the just obtained expression of $\mathbf{s}$, it results: $\mathbf{x} = \mathbf{As} = -\frac{1}{p}\mathbf{A}\mathbf{P}^{-1}(\mathbf{s})\mathbf{A}^T\boldsymbol{\alpha}$, or $\boldsymbol{\alpha} = -p\{\mathbf{A}\mathbf{P}^{-1}(\mathbf{s})\mathbf{A}^T\}^{-1}$;

27.3 Projection-Based sPGD

The main limitation of SSL is the cost when the number of parameters increase. As discussed before, when choosing the lowest level, the zero-level, we must consider two points per dimension, related to the minimum and the maximum parameter values. Thus, if we consider P parameters, the lowest approximation level, leading to a multilinear approximation, involves 2^P collocation points, that coincides with the number of high-fidelity solutions to be computed. This value becomes excessive when the number of parameters increases, for example, for P $= 10$, one must solve 2^{10} problems.

When considering the sPGD, the first advantage was the possibility of using any basis, in particular the possibility of defining constant, simply-linear (linear in one coordinate and constant in all the others), bi-linear (linear in two coordinates and constant in the others), and so on. This low-degree richness enabled cheap approximations of functions evolving smoothly within the parametric domain, where the original SSL failed because the drawback just discussed.

Both techniques, the SSL and the sPGD, were based on collocation, the former hierarchical, the latter sparse. However, projection methods offer many advantages when combined with quadratures. They combines accuracy and non-intrusivity.

A one-dimensional quadrature reads:

$$\int_{\mathcal{I}} f(x)\, dx = \sum_{i=1}^{M} \omega_i f(x_i),$$

where the points x_i and the weights ω_i are calculated for integrating exactly M monomials of a given polynomial basis. Some quadratures, like the one of Gauss, widely considered in finite element settings, with M points allows the exact integration of a polynomial of degree $2M - 1$. Multidimensional integrals are performed by simply considering product rules.

The most salient property of projection methods based on quadratures, is that the asymptotic accuracy of a quadrature rule is determined by the highest total degree N for which we can guarantee that all the monomials will be integrated exactly. For example, when addressing the 2D case with degree four, the 15 monomials illustrated in Fig. 27.2 (top) suffices to reach an asymptotic accuracy of degree four, without the necessity of including the remaining 10. Even if in 2D the saving is not spectacular, in 5 dimensions and degree 10, 252 monomials suffice, with 99748 useless monomials (with respect to a 10°C asymptotical accuracy).

The next ingredient contributing to the global efficiency is the possibility of considering a hierarchical quadrature. Smolyak quadrature (widely considered in sparse-grid based discretizations) provides a valuable alternative. Figure 27.2 (bottom) depicts the location of the 1D quadrature points when increasing the level, starting from one point, then 2, then adding a new point in each of the intervals resulting at the previous level.

Fig. 27.2 Quadrature in 2D and degree four (top) and 1D Smolyak quadrature (bottom)

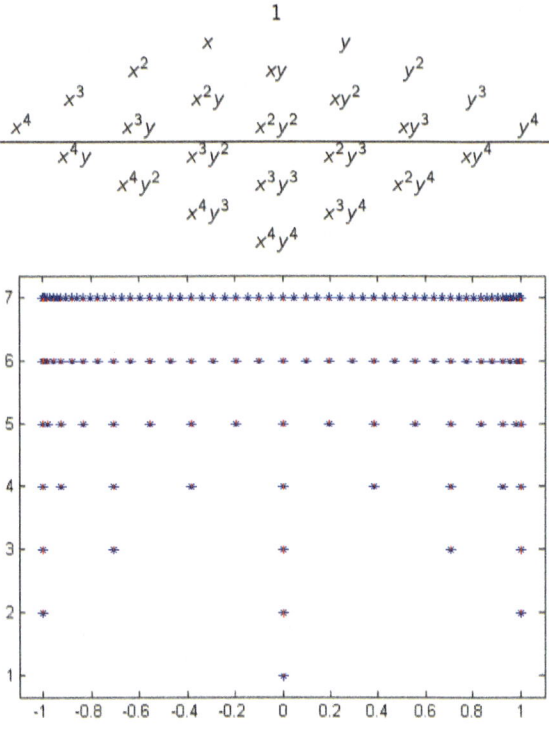

Projections need the calculation of integrals. In 1D, if one looks to approximate $f(x)$ in an appropriate functional basis $\mathcal{N}_i(x)$

$$f(x) \approx \sum_{i=1}^{N} \alpha_i \mathcal{N}_i(x),$$

the projection reads

$$\int \left(\sum_{i=1}^{N} \alpha_i \mathcal{N}_i(x) \right) \mathcal{N}_k(x) \, dx = \int f(x) \mathcal{N}_k(x) \, dx, \quad \forall k,$$

which when considering an orthonormal basis reduces to

$$\alpha_k = \int f(x) \mathcal{N}_k(x) \, dx.$$

Now by using the quadrature to evaluate the integral of the right-hand member, results in

$$\alpha_k = \sum_{j=1}^{M} \omega_j f(x_j),$$

which only requires the knowledge of $f(x)$ at the quadrature points.

The just described procedure can be extended to many dimensions, and it can be combined with the standard separated representation constructors (like in the SSL and sPGD), and only requires computing the solution of the high-fidelity models at the quadrature points (e.g. the ones of the Smolyak) with a product rule.

27.4 From ANOVA-Based Metamodelling to ANOVA-Based sPGD

From the definition of functions $f_s(\mathbf{x}_s)$ given in Sect. 15, and the alternative formulation that replaces expectations by their anchored counterparts, e.g.

$$f_{x_i}(x_i) = f(c_1, \ldots, c_{i-1}, x_i, c_{i+1}, \ldots, c_D) - f(\mathbf{c}),$$

the function $f(\mathbf{x})$ is expressed as

$$f(\mathbf{x}) = f(\mathbf{c}) + \sum_{i_1=1}^{D} \left(f(\mathbf{c}|x_{i_1}) - f(\mathbf{c}) \right) + \cdots$$

Thus, the functional approximation simply needs adding different regressions, taking into account that the higher order terms are expected to vanish (very) quickly.

Any linear or nonlinear regression can be used to approximate the functions involved in the previous decomposition, from the simplest ones depending on a single variable $f(\mathbf{c}|x_{i_1})$, to the ones involving many coordinates. The curse of dimensionality is avoided because as just mentioned, lowest order terms are expected sufficing to approximate the function [4, 7].

27.4.1 ANOVA-Based sPGD

If higher order terms are requested, one could imagine approximating these functions making use of the sparse PGD, from a few high-fidelity solutions computed at some points related to a sampling or a valuable quadrature.

The advantage of the ANOVA-based decomposition with respect to a purely sPGD is the statistical hierarchical decomposition which introduces all single-variables dependences, then the pair-correlations, and so on.

27.4.2 Statistics

As soon as the function decomposition is obtained, this surrogate can be used to compute the statistical moments using for example Monte-Carlo. Thus, the ANOVA decomposition can be used as a surrogate to quantify uncertainty.

27.5 Multi-PGD, mPGD

In order to better approach nonlinear manifolds, the PGD can be used as a locally linear (or low degree nonlinear) approximation [8, 9]. When solving a (parametric) partial differential equation (within the intrusive framework) special attention must be paid in order to ensure the continuity of the approximation. This constraint was addressed within the partition of unity (PU) method.

However, when using collocation techniques to generate parametric separated representations, the continuity is not a restriction anymore, and the previous techniques (SSL or sPGD) can be applied to different patches (disjoint or overlapping) covering the domain where the problem is defined. For the sake of simplicity but without any loss of generality, in what follows we consider a simple 2D problem.

27.5.1 Intrusive Framework

As just mentioned, within the intrusive framework, mPGD employs the Partition of Unity, PU, method. In essence, the PU method is based on the fact that, given a collection of patches defined over the domain Ω, $\Omega_i, i = 1, \ldots, n_{\text{patch}}$, a partition of unity φ_i defined on these patches (i.e., $\sum_i \varphi_i = 1$ everywhere in the domain), and function spaces V_i, associated to each patch, the approximation obtained by

$$V = \sum_{i=1}^{n_{\text{patch}}} \varphi_i V_i$$

inherits the approximation properties of the spaces V_i and the continuity properties of φ_i.

The finite element approximation, $u(x, y) = \sum_i N_i(x, y) u_i$ is a particular instance of the partition of unity, since it satisfies $\sum_i N_i(x, y) = 1$, and also linear consistency properties.

The main ingredient of the mPGD is the combination of the finite element shape functions, $N_i(x, y)$, which constitute a very convenient example of partition of unity, and the PGD as enrichment, i.e.

$$u(x, y) = \sum_{i \in \mathcal{I}} N_i(x, y) \sum_{k=1}^{N} X_k^i(x) Y_k^i(y),$$

27.5 Multi-PGD, mPGD

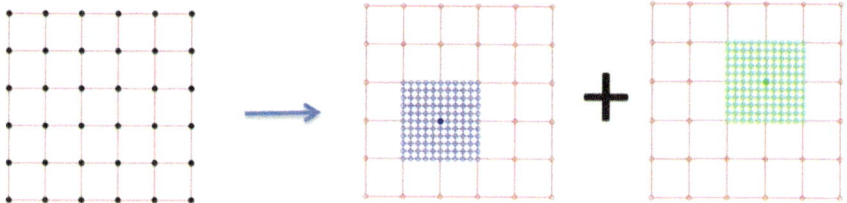

Fig. 27.3 (Left) piecewise linear partition of unity; (centre) domain of influence of the ith PGD; (right) domain of influence of the kth PGD

where $X_k^i(x)$ and $Y_k^i(x)$ functions are the k-th one-dimensional modes related to the i-th PGD, with $\mathcal{I}$ the set of PGD-enriched finite element supports.

To define the trial function, the simplest choice consists of the ones related to the Galerkin formulation, which when looking for the n-th PGD term reads

$$u^*(x, y) = \sum_{i \in \mathcal{I}} N_i(x, y) \left(X_n^{i*}(x) Y_n^i(y) + X_n^i(x) Y_n^{i*}(y) \right).$$

Depending on how finite element shape functions are defined, the properties of the method will change. Figure 27.3 sketches the partition when using piecewise linear functions with patches overlapping. Note how a quadrilateral element has contributions coming from four different PGDs. This kind of partition ensures a smooth transition between PGD approximations, which may vary continuously along the domain in accordance with the partition of the domain.

27.5.2 Non-intrusive Framework

Performing a single regression valid for a large physical and parametric domain is often a difficult task. On one hand, the construction of a regression for a quantity of interest, QoI, is in general more accurate than obtaining the time evolution of the solution at a certain point. This is, in turn, more accurate than creating a regression for a field variable. The origin of this problem can be traced back to the fact that, in general, regressions are constructed by using the L2-norm. Therefore, if a given field exhibits strong localizations, these local features are in general smoothed out in benefit of an averaged solution, good on an average sense.

To improve accuracy, an appealing choice consists of the partition the physical space, so as to perform a regression within each of the resulting patches. Local quasi-linear regressions perform in general better than rich nonlinear and global regressions in the whole spatial domain. However, continuity can be lost on the patch boundaries when using multiple regressions, one per patch. To enforce continuity, it is possible, for instance, to resort to the Partition of Unity methods. However, as continuity is not mandatory when operating from surrogates, in the vicinity of patch boundaries,

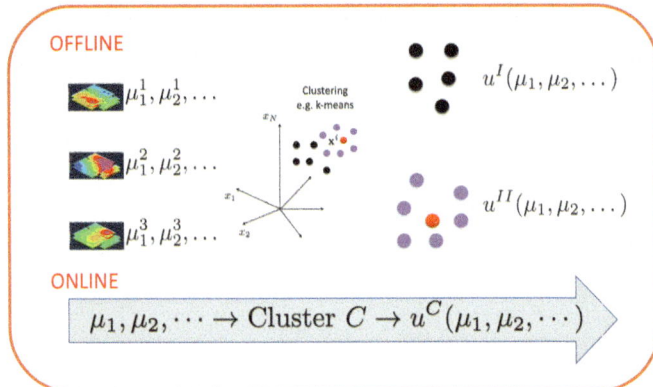

Fig. 27.4 Multi-PGD: (i) Clustering high-fidelity solutions related to a design of experiments; (ii) Create a regression in each cluster; and (iii) Constructing a classifier able to associate a cluster to any parameters choice

one could compute regressions covering both sides of the boundaries and average them. Another alternative is taking advantage of these discontinuities for refinement purposes. This is usually done in the finite element method framework.

This is also valid not only for the spatial domain, but also for the parametric one. In that last case, making a partition of the (possibly high-dimensional) multi-parametric space is not evident. One possibility consists in clustering solutions related to the considered sampling, for example by employing *k-means*. Then, a nonlinear regression is created from the solutions for each cluster. Finally, the trickiest issue becomes the way of associating a cluster to any choice of parameters. In other words, performing an accurate classification. The global picture of this last procedure is depicted in Fig. 27.4.

27.6 Parametric Transfer Function, PTF

The parametric transfer function can be constructed by parametrizing the model inputs (loading). This task is not always easy because the inputs can involve too many degrees of freedom.

Imagine for a while that one is interested parametrizing the initial state of a solid in elastodynamics. Thus, a priori, a complete parametrization will consist of the displacement and the velocity at each node of the mesh attached to the domain Ω. This choice is robust but extremely computationally expensive. To enhance computational efficiency, the simplest route consists in extracting a reduced basis of the displacement and velocities by considering different snapshots and applying POD.

As soon as the reduced basis representing the problem has been obtained, any input can be expanded in it, according to

$$\mathbf{u}(\mathbf{x}) = \sum_{i=1}^{N_u} \alpha_i \boldsymbol{\phi}_i^u,$$

and

$$\dot{\mathbf{u}}(\mathbf{x}) = \sum_{j=1}^{N_v} \beta_i \boldsymbol{\phi}_j^v.$$

Now, the standard constructors (intrusive or non-intrusive) apply, to compute the general parametric solution $u(\mathbf{x}, \alpha_1, \ldots, \alpha_{N_u}, \beta_1, \ldots, \beta_{N_v}, \boldsymbol{\mu})$ which constitutes the parametric transfer function, PTF.

References

1. D. Borzacchiello, J.V. Aguado, F. Chinesta, Reduced Order Modelling for efficient process optimisation of a hot-wall chemical vapour deposition reactor. Int. J. Numeri. Methods Heat Fluid Flow **27**(7), 1602–1622 (2017)
2. D. Borzacchiello, J.V. Aguado, F. Chinesta, Non-intrusive sparse subspace learning for parametrized problems. Arch. Comput. Methods Eng. **26**(2), 303–326 (2019)
3. R. Ibanez, E. Abisset-Chavanne, A. Ammar, D. Gonzalez, E. Cueto, A. Huerta, J.L. Duval, F. Chinesta, A multi-dimensional data-driven sparse identification technique: the sparse proper generalized decomposition. Complexity, Article ID 5608286 (2018)
4. A. Sancarlos, V. Champaney, E. Cueto, F. Chinesta, Regularized regressions for parametric models based on separated representations. Adv. Model. and Simul. in Eng. Sci. **10**(4) (2023). https://doi.org/10.1186/s40323-023-00240-4
5. S. Brunton, J.L. Proctor, N. Kutz, Discovering governing equations from data by sparse identification of nonlinear dynamical systems. PNAS **113**(15), 3932–3937 (2016)
6. Z. He, S. Xie, On the convergence of FOCUSS algorithm for sparse representation. IEEE Trans. Inform. Theory (2012)
7. M. Kubicek, E. Minisci, M. Cisternino, High dimensional sensitivity analysis using surrogate modeling and high dimensional model representation. Int. J. Uncertainty Quantification **5**(5), 393–414 (2015)
8. A. Badias, D. Gonzalez, I. Alfaro, F. Chinesta, E. Cueto, Local Proper Generalized Decomposition. Int. J. Numer. Methods Eng. **112**(12), 1715–1732 (2017)
9. R. Ibanez, E. Abisset-Chavanne, F. Chinesta, A. Huerta, E. Cueto, A local, multiple Proper Generalized Decomposition based on the Partition of Unity. Int. J. Numer. Methods Eng. **120**(2), 139–152 (2019)

Open Access This chapter is licensed under the terms of the Creative Commons Attribution-NonCommercial-NoDerivatives 4.0 International License (http://creativecommons.org/licenses/by-nc-nd/4.0/), which permits any noncommercial use, sharing, distribution and reproduction in any medium or format, as long as you give appropriate credit to the original author(s) and the source, provide a link to the Creative Commons license and indicate if you modified the licensed material. You do not have permission under this license to share adapted material derived from this chapter or parts of it.

The images or other third party material in this chapter are included in the chapter's Creative Commons license, unless indicated otherwise in a credit line to the material. If material is not included in the chapter's Creative Commons license and your intended use is not permitted by statutory regulation or exceeds the permitted use, you will need to obtain permission directly from the copyright holder.

Chapter 28
Quantities Of Interest

An engineer is more concerned by Quantities of Interest, QoI, than by the problem solution itself. The problem solution is very rich, it provides the value of the field of interest everywhere, at any time, and within a parametric framework, for any choice of the model parameters. However, designs must optimize some criteria, minimize some cost function or ensure the fulfillment of some constraints, all of which are related explicitly or implicitly to the problem solution. This section aims at deriving a parametric expression of a QoI, $\mathcal{O}$, depending on the problem's solution $u(\mathbf{x}, t, \boldsymbol{\mu})$.

28.1 Parametric QoI

The previous sections addressed the construction of a parametric solution by using both intrusive and non-intrusive techniques. Thus, the generic solution of a parametrized PDE is assumed to be expressed in the separated form:

$$u(\mathbf{x}, t, \boldsymbol{\mu}) = u(\mathbf{x}, t, \mu_1, \ldots, \mu_P) \approx \sum_{i=1}^{N} X_i(\mathbf{x}) T_i(t) M_i^1(\mu_1) \cdots M_i^P(\mu_P),$$

or in its equivalent discrete tensor form

$$\mathbf{U} \approx \sum_{i=1}^{N} \mathbf{X}_i \otimes \mathbf{T}_i \otimes \mathbf{M}_i^1 \otimes \cdots \otimes \mathbf{M}_i^P,$$

with $\mathbf{U}$ the multi-dimensional tensor whose $\{k, l, m_1, \ldots, m_P\}$-th entry contains the value of the field u at point, time and parameters referred by these indexes, i.e., $u(\mathbf{x}_k, t_l, \mu_{1_{m_1}}, \ldots, \mu_{P_{m_P}})$. Obviously, at any other point different to a node of the spatial mesh $(\mathbf{x}_k)$, time (t_l) or parameters $(\mu_{1_{m_1}}, \cdots)$, the solution is computed simly by interpolation.

© The Author(s) 2025
F. Chinesta et al., *A Gentle Introduction to Data, Learning, and Model Order Reduction*,
Studies in Big Data 174, https://doi.org/10.1007/978-3-031-87572-4_28

We assume the QoI related to the problem's solution, expressed by $\mathcal{O}(\mathbf{x}, t, \boldsymbol{\mu})$. Thus, we can compute the output at the collocation points when using the SSL technique or in the points of a sparse sampling (e.g., the ones associated to the Latin Hypercube sampling), $\mathcal{O}_m$, with $m = 1, \ldots, M$, referring to a particular choice of the problem coordinates (space, time and parameters). Now, using the regressions previously discussed the parametric QoI can be derived

$$\mathcal{O}(\mathbf{x}, t, \boldsymbol{\mu}) \approx \sum_{i=1}^{O} \mathcal{X}_i(\mathbf{x}) \mathcal{T}_i(t) \mathcal{M}_i^1(\mu_1) \cdots \mathcal{M}_i^P(\mu_P).$$

To compute the sensitivity of the output to a given parameter μ_1, one proceeds by simply compute

$$\frac{\partial \mathcal{O}(\mathbf{x}, t, \boldsymbol{\mu})}{\partial \mu_1} \approx \sum_{i=1}^{O} \mathcal{M}_i(\mathbf{x}) \mathcal{T}_i(t) \frac{\partial \mathcal{M}_i^1(\mu_1)}{\partial \mu_1} \mathcal{M}_i^2(\mu_2) \cdots \mathcal{M}_i^P(\mu_P),$$

that offers an explicit expression of sensitivities. These expressions are of primary interest in optimization problems.

28.2 Uncertainty Quantification

The parameters joint probability distribution function can be expressed in the separated form

$$\Xi(\mu_1, \ldots, \mu_P) \approx \sum_{i=1}^{S} \mathcal{F}_i^1(\mu_1) \cdots \mathcal{F}_i^P(\mu_P),$$

that reduces to the uncorrelated case for $S = 1$.

By combining it with the parametric QoI

$$\mathcal{O}(\mathbf{x}, t, \boldsymbol{\mu}) \approx \sum_{i=1}^{O} \mathcal{M}_i(\mathbf{x}) \mathcal{T}_i(t) \mathcal{M}_i^1(\mu_1) \cdots \mathcal{M}_i^P(\mu_P),$$

the calculation of the different statistical moments becomes straightforward. Thus, the first moment, the average field, reads

$$\overline{\mathcal{O}}(\mathbf{x}, t) = \int_{\Omega_1 \times \cdots \times \Omega_P} \mathcal{O}(\mathbf{x}, t, \mu_1, \ldots, \mu_P) \, \Xi(\mu_1, \ldots, \mu_P) \, d\mu_1 \cdots d\mu_P,$$

where Ω_k denotes the domain of parameter μ_k. The separated representation is a key point for the efficient evaluation of this multidimensional integral, which becomes a series of one dimensional integrals. The calculation of higher order statistical moments (variance, ...), marginal or conditional probabilities, proceeds in a similar manner [1].

28.3 Adjoint Method

Optimization procedures in parametric settings can be performed by considering a variety of techniques. In our works, we extensively employed surrogate models (parametric regressions) for that purpose, combined with any deterministic (gradient-based) or stochastic (e.g. genetic algorithm) optimization procedure. As just discussed the fact of having an explicit expression of the solution sensitivity becomes very valuable for making optimization.

When the parametric solution is not available, the so-called *Adjoint* method can facilitate the task. For the sake of completeness, we summarizes the main ingredients of Adjoint methods in what follows.

We assume the solution of a parametrized problem as $F(u, p) = 0$, and a quantity of interest, QoI, expressed by $\mathcal{O} = \mathcal{G}(u)$, to be optimized with respect to the parameters choice. Thus, one is looking for the parameter choice p that allows maximizing $\mathcal{O}$.

For that purpose, the quantities $\frac{\partial \mathcal{G}}{\partial p}$ are of major interest for driving the maximization process.

As $\mathcal{G}$ does not depends explicitly on p, its gradient can be computed from

$$\frac{\partial \mathcal{G}}{\partial p} = \frac{\partial \mathcal{G}}{\partial u} \frac{\partial u}{\partial p}. \tag{28.1}$$

When the explicit form of $\mathcal{G}(u)$ is available (it is in general the case), as well as the one of $u(p)$ (as it is the case when operating with the PGD) the just referred quantities (28.1) can be immediately calculated.

When it is not the case, that is, when the parametric solution $u(p)$ is not explicitly available, having the implicit form $F(u, p) = 0$, the Adjoint method applies. The fact that $\frac{dF}{dp} = 0$, leads to

$$\frac{dF}{dp} = \frac{\partial F}{\partial u} \frac{\partial u}{\partial p} + \frac{\partial F}{\partial p} = 0,$$

or

$$\frac{\partial u}{\partial p} = -\left(\frac{\partial F}{\partial u}\right)^{-1} \frac{\partial F}{\partial p}.$$

Thus,

$$\frac{\partial \mathcal{G}}{\partial p} = -\frac{\partial \mathcal{G}}{\partial u} \left(\frac{\partial F}{\partial u}\right)^{-1} \frac{\partial F}{\partial p} = -\left(\frac{\partial \mathcal{G}}{\partial u} \left(\frac{\partial F}{\partial u}\right)^{-1}\right) \frac{\partial F}{\partial p} = -\lambda \frac{\partial F}{\partial p},$$

where λ results from the solution of the so-called adjoint problem

$$\frac{\partial F}{\partial u}\lambda = \frac{\partial \mathcal{G}}{\partial u}.$$

Reference

1. V. Limousin, X. Delgerie, E. Leroy, R. Ibanez, C. Argerich, F. Daim, J.L. Duval, F. Chinesta, Advanced model order reduction and artificial intelligence techniques empowering advanced structural mechanics simulations. Application to crash test analyses. Mech. Indus. **20**(8), Article 804 (2019)

Open Access This chapter is licensed under the terms of the Creative Commons Attribution-NonCommercial-NoDerivatives 4.0 International License (http://creativecommons.org/licenses/by-nc-nd/4.0/), which permits any noncommercial use, sharing, distribution and reproduction in any medium or format, as long as you give appropriate credit to the original author(s) and the source, provide a link to the Creative Commons license and indicate if you modified the licensed material. You do not have permission under this license to share adapted material derived from this chapter or parts of it.

The images or other third party material in this chapter are included in the chapter's Creative Commons license, unless indicated otherwise in a credit line to the material. If material is not included in the chapter's Creative Commons license and your intended use is not permitted by statutory regulation or exceeds the permitted use, you will need to obtain permission directly from the copyright holder.

Chapter 29
Projection and Transport

When the model parameters concern the geometry of the domain in which different fields are defined, the parametric interpolation-based techniques described in the present monography are faced with the difficulty of having different geometries supporting the parametric fields.

The simplest solution consists in defining a parent domain, and projecting all the geometries with their attached fields into that reference domain. Once all the fields related to the different choices of the parameters are expressed on the same domain, interpolation proceeds from its most standard way.

The main point is the way of performing the mapping from a general geometry to the reference one, e.g., a general 2D surface embedded in $\mathbb{R}^3$ to the reference domain that consists of a plate domain in $\mathbb{R}^2$, while trying to keep unaltered all the important features present in the solutions during the mapping process, to avoid spurious numerical artifacts.

29.1 Riemannian Projectors

The just referred problem has been widely addressed in the domain of computational geometry, and from the different existing solutions, conformal mappings are the ones that we employed in our former works [1]. The process of mapping while enforcing the angles preservation constitutes a very appealing alternative. One technique able to perform in that sense consists in applying the Ricci flow or its variant, the Yamabe flow. The interested reader can refer to [2–6].

The Ricci flow is formulated as follows: Given a smooth manifold $\mathcal{M}$ and an interval $\mathcal{I}$, the Ricci flow computes at $t \in \mathcal{I}$ the Riemannian metric $g(t)$ on $\mathcal{M}$, by solving

$$\frac{\partial g}{\partial t} = -2\mathrm{Ric}.$$

Different forms exist to express the right-hand member. In the case of constant curvature and Einstein metric, Ric $= \lambda g$, with $g(t = 0) = g_0$ from which it results:

- If $\lambda > 0$, the Ricci flow contracts g as time progresses;
- If $\lambda = 0$, the initial Ricci-flat is stationary, without constraint for the time;
- If $\lambda < 0$, the Ricci flow expands g, and the Ricci flow is said *immortal*.

In general the curvature is non constant, and Ric $=$ **G**, with **G** the Ricci curvature tensor. As this tensor implies second derivatives of g, formally this equations resembles the heat equation, and in consequence it tends to reduce the curvature at local maxima, and increase it at local minima, exactly like the heat transfer equation produces on temperature fields, trying to make it uniform.

29.1.1 Ricci Flow Based Conformal Mapping and Its Associated Reduction Procedure

Thus, Ricci flow can be seen as a metric diffusion, with consequently several appealing properties to flatten manifolds (surfaces embedded in $\mathbb{R}^3$) while keeping the most appealing properties of the metrics: (i) Gauss curvature during surface Ricci flow is bounded, thereby ensuring the numerical stability; (ii) Surface Ricci flow is conformal, that is, the deformation of the Riemannian metric preserves the angles; (iii) It provides intuitive geometric interpretations; and (iv) Ricci flow is variational and hence the surface Ricci flow can be formulated as a convex optimization problem which has a unique global optimum and can be solved efficiently using the Newton method.

Thus the reduction procedure consists of: (i) flatten source and target surfaces; (ii) re-scale the flattened source mesh and rotate to align with the target one; (iii) find the location of the nodes in the source mesh, on the target one; and (iv) retrieve the morphed mesh. For more details the interested reader car refer to [1].

29.2 Smart Mapping

To visualize the limitation of usual interpolation, in what follows we consider an elastic rope fixed at its two ends, A and B, with an enforced unit displacement at point P_i of coordinate $x = X$, as depicted in Fig. 29.1. The associated top deformed configuration is assumed linear between each support and the applied displacement.

Applying a unit displacements at P_1 and another one at position P_2, would provide us with two different deformations, shown in Fig. 29.2. If now we consider the unit displacement occuring at the middle point between P_1 and P_2, classical interpolation of the two displacement fields leads to the solution depicted by the red curve in Fig. 29.3, even if as expected, the right solution in this case consists in the green curve shown in Fig. 29.3.

29.2 Smart Mapping

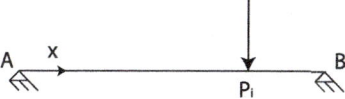

Fig. 29.1 An elastic rope that will experience a displacement at position $x = P_i$ as depicted in Fig. 29.2

(a) Displacement enforced at position P_1 (b) Displacement enforced at position P_2

Fig. 29.2 Solution for two different applied displacements

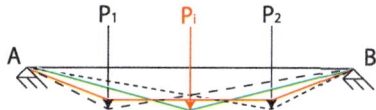

Fig. 29.3 Interpolation of the displacement field using the standard interpolation in red, and the reference solution in green

The problem of the deformation of the rope illustrated in Fig. 29.2 can be formulated as a 2D problem defined in domain (x, X) where x is the coordinate along the length of the rope $x \in [0, L]$, with L the length of the undeformed rope (that is, the distance between both supports, A and B), and X the position where the displacement occurs, such that $X \in [a, b]$ (where a tends to 0 and b to L).

The exact solution of the vertical displacement $v(x)$ depends x and X, reads:

$$v(x; X) = \begin{cases} -\frac{x}{X} & \text{if } x \leq X \\ -\frac{L-x}{L-X} & \text{if } x \geq X \end{cases}.$$

By using a geometrical transformation, we can place the point of application of the displacements P_i at the same location, for example, at the middle point $L/2$. Thus, the coordinate $x \in [0, L]$ is transformed into $s \in [0, 2]$ according to

$$x = \begin{cases} sX & \text{if } s \in [0, 1] \\ (s-1)(L-X) + X & \text{if } s \in [1, 2] \end{cases}.$$

Using this transformation the load is always applied at $s = 1$, and therefore the solution in the (s, X) domain becomes trivially separable. This efficient mapping was widely analyzed in [7, 8].

29.3 From Optimal Transport to Parametric Optimal Transport

Optimal transport (OT) problem was introduced by Monge to calculate the optimal way to transport a certain amount of soil from an initial landscape to a desired target landscape. The considered cost (to be minimized) in that case was the traveled distance. The Monge's problem can be illustrated by considering a resource produced by N mines and transported to M factories while minimizes the square of the traveled Euclidean distance.

Each mine $n \in [\![N]\!]$ (the notation $[\![N]\!]$ refers to $[1, \ldots, N]$), located at x_n, produces a quantity a_n and each factory $m \in [\![M]\!]$, located at y_m, consumes a quantity b_m. Thus, the produced and consumed distributions, respectively α and β can be expressed from

$$\alpha = \sum_{n=1}^{N} a_n \delta_{x_n} \quad \text{and} \quad \beta = \sum_{m=1}^{M} b_m \delta_{y_m}.$$

The Monge problem looks for the map T that connects each point x_n with a single point y_m such that it pushes α toward β. Since no resource can be produced or destroyed during the transport, the map $T : \{x_1, \ldots, x_N\} \to \{y_1, \ldots, y_M\}$ must verify the mass conservation

$$\forall m \in [\![M]\!], b_m = \sum_{n:T(x_n)=y_m} a_n,$$

that can be expressed in the compact form $T_{\#}\alpha = \beta$.

It can be noted that this map is surjective. Moreover, this map must minimized the cost defined as the square of the Euclidian distance between the sources and their associated destinations:

$$C_{x_n, y_m} = \|x_n - y_m\|_2^2,$$

resulting in the minimization problem:

$$\min_{T} \sum_{n=1}^{N} C_{x_n, T(x_n)}. \tag{29.1}$$

The problem can be simplified by assuming the same number of mines and factories, i.e. $N = M$, and that each mine produces the same amount of resource and each factory consumes this same amount of resource, i.e. $a_n = b_m = 1/N$. In these circumstances, the map T is a bijection and the the minimization problem, Eq. (29.1), becomes a deterministic assignment problem.

By considering the just introduced hypotheses, the 1D Monge problem illustrated in Fig. 29.4 (equivalent to the 1D optimal matching problem) represents the simplest

29.3 From Optimal Transport to Parametric Optimal Transport

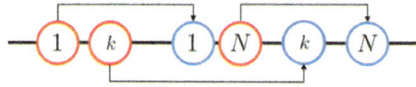

Fig. 29.4 1D Monge problem with $N = M$ and $a_n = b_m = 1/N$. Mines are represented by red circles and factories by blue ones. The optimal matching is illustrated by black arrows

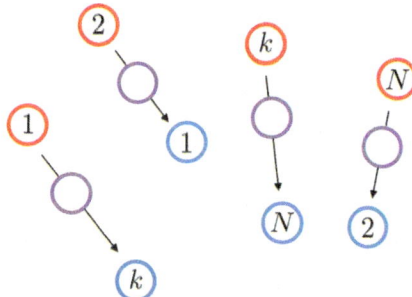

Fig. 29.5 2D Monge problem with $N = M$ and $a_n = b_m = 1/N$. Mines are represented by red circles and factories by blue ones. The optimal matching is illustrated by black arrows. The interpolated distribution is illustrated by violet points

Optimal Transport problem. In this case the solution simply consists in sorting the mines and factories locations along the 1D axis in ascending order, as illustrated in the figure, where the mines are represented by red circles and the factories by blue ones.

The 2D problem becomes much more complex since no order relation exists, but remains easy to solve using, for instance, linear programming. Indeed, the problem consists of the optimal assignment problem between two point clouds (assuming each cloud composed by the same number of points, each having the same amount of mass and the cost given by the square of the Euclidian distance), as illustrated in Fig. 29.5 where each point has a x and y coordinates. Again, mines are represented by the red circles and factories by the blue ones.

Once both point clouds are optimally matched, it is possible to optimally interpolate, between the two distributions by partially displacing all the points over the corresponding segments. The resulting interpolated distribution (or point cloud) is illustrated in Fig. 29.5 (violet circles).

The construction of the parametric optimal transport (POT) model follows four steps. First, all the distributions, among which it is intended to interpolate, are decomposed into a sum of Gaussians. Then, each Gaussian from each distribution is paired with one, and only one, Gaussian from every other distribution using an adequate technique, for example, a Genetic Algorithm. Next, a POD is applied over the positions of all the Gaussians of each distribution. Finally, a regression is applied over the coefficients of this POD leading to a parametric optimal transport based model which can be accessed in an online manner. For more details in the implementation the interested reader can refer to [9].

References

1. A. Ebey-Thomas, S. Guevelou, E. Di Pasquale, A. Chambard, J.L. Duval, F. Chinesta, V. Limousin, X. Delgerie, E. Leroy, Shape parametrization & morphing in sheet-metal forming. Procedia Manuf. **47**, 702–706 (2020)
2. F. Luo, Combinatorial Yamabe flow on surfaces. Commun Contemp. Math. **6**, 765–780 (2004)
3. M. Jin, J. Kim, X.D. Gu, Discrete surface Ricci flow: theory and applications, in *Mathematics of Surfaces XII. Mathematics of Surfaces 2007. Lecture Notes in Computer Science*, vol 4647, eds. by R. Martin, M. Sabin, J. Winkler (2007)
4. X.D. Gu, F. Luo, S.T. Yau, Recent advances in computational conformal geometry, in *Mathematics of Surfaces XIII*, eds.by E.R. Hancock, R.R. Martin, M.A. Sabin (2009), pp. 189–221
5. W. Zeng, D. Samaras, X.D. Gu, Ricci flow for 3d shape analysis. IEEE Trans Pattern Anal. Mach. Intel. **32**, 662–677 (2010)
6. W. Zeng, X.D. Gu, *Ricci Flow for Shape Analysis and Surface Registration* (Springer, New York, 2013)
7. A. Ammar, Ch. Ghnatios, F. Delplace, A. Barasinski, J.L. Duval, E. Cueto, F. Chinesta, On the effective conductivity and the apparent viscosity of a thin-rough polymer interface using PGD-based separated representations. Int. J. Numer. Methods Eng. **121**(23), 5256–5274 (2020)
8. C. Ghnatios, E. Cueto, A. Falco, J.L. Duval, F. Chinesta, Spurious-free interpolations for non-intrusive PGD-based parametric solutions: application to composites forming processes. Int. J. Mater. Forming (2020)
9. S. Torregrosa, V. Champaney, A. Ammar, V. Hebert, F. Chinesta, Surrogate parametric metamodel based on optimal transport. Math. Comput. Simul. **194**, 36–63 (2022)

Open Access This chapter is licensed under the terms of the Creative Commons Attribution-NonCommercial-NoDerivatives 4.0 International License (http://creativecommons.org/licenses/by-nc-nd/4.0/), which permits any noncommercial use, sharing, distribution and reproduction in any medium or format, as long as you give appropriate credit to the original author(s) and the source, provide a link to the Creative Commons license and indicate if you modified the licensed material. You do not have permission under this license to share adapted material derived from this chapter or parts of it.

The images or other third party material in this chapter are included in the chapter's Creative Commons license, unless indicated otherwise in a credit line to the material. If material is not included in the chapter's Creative Commons license and your intended use is not permitted by statutory regulation or exceeds the permitted use, you will need to obtain permission directly from the copyright holder.

Chapter 30
Multiscale

Problems exhibiting multi-scale behaviors can be again addressed using separated representations, by adding different extra-dimensions (coordinates) to capture each scale present in the problem solution.

30.1 Partition of Unity-Based Enrichment

Again, the procedure proposed here makes use of the Partition of Unity. For the sake of simplicity the procedure will be described in a one-dimensional two-scale problem.

Thus, if function $u(x)$ is approximated in the standard way from (e.g., FEM) from

$$u(x) = \sum_{i=1}^{N} u_i N_i(x),$$

if the solution contains many scales, the mesh has to capture the behavior associated to the finest scale, with the consequent impact on the mesh size, i.e. $N \gg 1$.

An alternative approximation reads

$$u(x) = \sum_{i=1}^{N} N_i(x) u_i \sum_{j=1}^{J} G_j(\tau(x - x_i)) g_j,$$

where x_i are the nodes in the macro-mesh, $\tau(x - x_i)$ is a dependent variable which presents an offset based on x_i, $G_j(\tau)$ is the j-th micro-scale approximation function and g_j its associated micro-scale degree of freedom.

Figure 30.1 shows the shape functions associated to both the macro-scale (top) and the micro-scale (bottom). As it can be seen, a two scale approach presents two

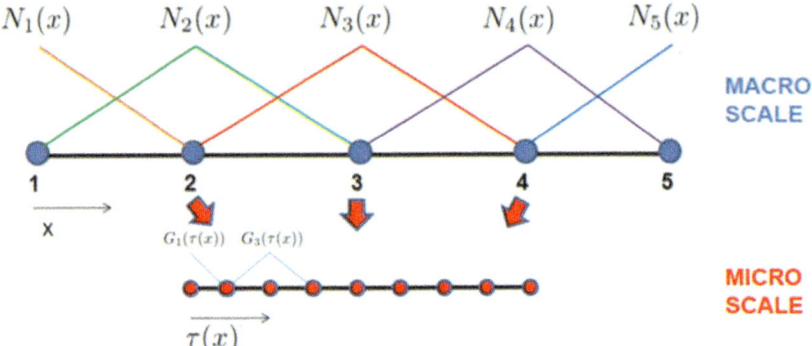

Fig. 30.1 Shape functions of a multi-scale approach: (top) macro shape functions; (bottom) micro shape functions

meshes related to micro and macro scales, respectively, with both scales remaining fully interconnected.

Moreover, it is worth mentioning the possibility of rewriting our solution function as $u(x, \tau)$, making it suitable for a PGD-like algorithm that proceeds from a separation of variables.

Therefore, following standard PGD rationale, the solution is sought in a greedy manner by means of a sum of M enrichments:

$$u(x, \tau) = \sum_{m=1}^{M} \sum_{i=1}^{N} N_i(x) u_i^m \sum_{j=1}^{J} G_j(\tau(x - x_i)) g_j^m.$$

30.2 Direct Separated Decompositions

The just discussed approximation proposed in [1] making use of the PU (partition of unity), mimics our proposal in [2], that was improved in [3], illustrated in Figs. 30.2 and 30.3. It generalizes the technique proposed in [4].

In [2, 3] the time axis (t) was cut, and after rotation if became 2D (t, τ), with the slow and fast times in the horizontal and vertical axis respectively. Figure 30.3 progresses in the cutting process, by adding finer and finer times-axes. The separation of variables at the heart of the PGD, allows solving a N-dimensional problem as a sequence of N one-dimensional problems solution. Thus, a mesh of 10^6 in 1D, becomes a $10^3 \times 10^3$ in 2D, and using the PGD its solution reduced to few 1D problems of size 10^3. Thus, by moving from 1D to 2D and using the PGD the complexity reduced from 10^6 to 10^3. If we move to $3D$, the complexity reduced to 10^2, $(10^2)^3 = 10^6$, and so on. Obviously there is a compromise, because using more and more dimensions could affect the convergence.

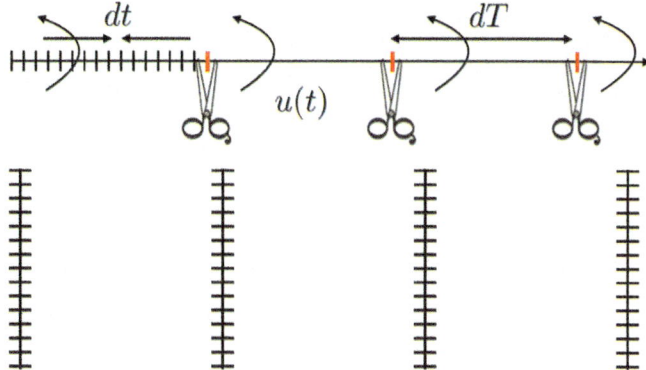

Fig. 30.2 Transforming a 1D problem into a 2D

Fig. 30.3 Adding more dimension for capturing finer scales

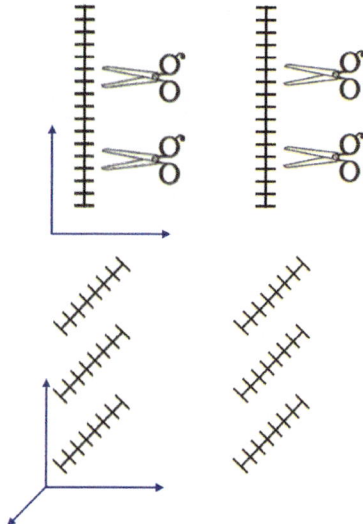

References

1. R. Ibanez, A. Ammar, E. Cueto, A. Huerta, J.L. Duval, F. Chinesta, Multi scale proper generalized decomposition based on the partition of unity. Int. J. Numer. Methods Eng. **120**(6), 727–747 (2019)
2. A. Ammar, F. Chinesta, E. Cueto, Coupling finite elements and proper generalized decompositions. Int. J. Multiscale Comput. Eng. **9**(1), 17–33 (2011)

3. A. Pasquale, A. Ammar, A. Falco, S. Perotto, E. Cueto, J.L. Duval, F. Chinesta, A separated representation involving multiple time scales within the proper generalized decomposition framework. Adv. Model. Simul. Eng. Sci. **8**, 26 (2021)
4. A. Badias, D. Gonzalez, I. Alfaro, F. Chinesta, E. Cueto, Local proper generalized decomposition. Int. J. Numer. Methods Eng. **112**(12), 1715–1732 (2017)

Open Access This chapter is licensed under the terms of the Creative Commons Attribution-NonCommercial-NoDerivatives 4.0 International License (http://creativecommons.org/licenses/by-nc-nd/4.0/), which permits any noncommercial use, sharing, distribution and reproduction in any medium or format, as long as you give appropriate credit to the original author(s) and the source, provide a link to the Creative Commons license and indicate if you modified the licensed material. You do not have permission under this license to share adapted material derived from this chapter or parts of it.

The images or other third party material in this chapter are included in the chapter's Creative Commons license, unless indicated otherwise in a credit line to the material. If material is not included in the chapter's Creative Commons license and your intended use is not permitted by statutory regulation or exceeds the permitted use, you will need to obtain permission directly from the copyright holder.

Chapter 31
Dynamics

Integrating dynamical systems remains expensive, and model order reduction seems an excellent gateway for making that integrations more efficient. We are discussing in this section some reduction approaches.

31.1 Transient Dynamics in the Physical Space

Structural solid dynamics is usually formulated either in the time or in the frequency domains. In the time domain, the general semi-discretized form reads

$$\mathbf{M}\frac{d^2\mathbf{U}(t)}{dt^2} + \mathbf{C}\frac{d\mathbf{U}(t)}{dt} + \mathbf{K}\mathbf{U}(t) = \mathbf{F}(t),$$

where $\mathbf{M}$, $\mathbf{C}$ and $\mathbf{K}$ are respectively the mass, damping and stiffness matrices, $\mathbf{U}$ the vector that contains the nodal displacements and $\mathbf{F}$ the nodal excitations (forces), of size N.

This formulation consists of N second order coupled ordinary differential equations, that can be integrated using either explicit, implicit or semi-implicit time discretizations. Explicit discretizations enable fast simulations, but the stability needs an adequate time-step, that decreases with the size of the elements employed in the space discretization.

Space-time separated representations where proposed in [1, 2] for enhancing integration performances, whereas in [3] a transfer function based real time integration within the PGD framework was successfully performed.

31.2 Mass Lumping and Modal Analysis

Usually, explicit integrations are combined with mass-lumping to recover a diagonal mass matrix, from which the inversion procedure becomes trivial, improving spectacularly the computational performances.

Modal analysis offers an appealing alternative route. It is based on the fact that the properties of the mass and stiffness matrices, imply the existence of a basis $\{\boldsymbol{\phi}_1, \boldsymbol{\phi}_2, \ldots, \boldsymbol{\phi}_N\}$, solution of the eigenproblem $\mathbf{K}\boldsymbol{\phi} = \omega^2 \mathbf{M}\boldsymbol{\phi}$, ensuring:

$$\boldsymbol{\phi}_i^T \mathbf{M} \boldsymbol{\phi}_j = \delta_{ij},$$

with δ the Kronecker delta; and

$$\boldsymbol{\phi}_i^T \mathbf{K} \boldsymbol{\phi}_j = \kappa_i \delta_{ij}.$$

By placing the eigenvectors $\boldsymbol{\phi}_i$ into the columns of matrix $\mathbf{P}$, the two previous relations can be rewritten as

$$\mathbf{P}^T \mathbf{M} \mathbf{P} = \mathbb{M} \equiv \mathbf{I},$$

and

$$\mathbf{P}^T \mathbf{K} \mathbf{P} = \mathbb{K},$$

$\mathbf{I}$ being the identity matrix, and $\mathbb{K}$ the diagonal matrix with entries $\mathbb{K}_{ii} = \kappa_i$.

Thus, if $\mathbf{C} = \mathbf{0}$ (undamped dynamics) or $\mathbf{C} = \alpha \mathbf{M} + \beta \mathbf{K}$ (proportional damping), using the coordinates $\boldsymbol{\varphi}$, $\mathbf{U} = \mathbf{P}\boldsymbol{\varphi}$, and pre-multiplying by $\mathbf{P}^T$, the dynamical problem reads

$$\mathbb{M} \frac{d^2 \boldsymbol{\varphi}(t)}{dt^2} + \mathbb{C} \frac{d\boldsymbol{\varphi}(t)}{dt} + \mathbb{K} \boldsymbol{\varphi}(t) = \mathbf{P}^T \mathbf{F}(t),$$

with $\mathbb{C}$ diagonal, i.e. $\mathbb{C}_{ij} = c_i \delta_{ij}$ ($\mathbb{C} = \mathbf{0}$ in the undamped case), that constitutes a system of N uncoupled second order ordinary differential equations that can be integrated very efficiently to provide transient responses.

Modal bases associated to different model parameter choices can be, first (i) classified using appropriate metrics (e.g., TDA in [4]); and (ii) then interpolated by using an adequate transport (e.g., Grassmann). The solution of parametric eigenproblems constitutes an active domain of research, in both an intrusive and a non-intrusive formulation.

31.3 Harmonic Analysis

Another alternative route consists in considering the frequency instead of the time as coordinate, that is, applying the Fourier transform to the original problem, very convenient when proceeding far from the transient regime, that is, in the forced regime, or when considering more general damping models. However, such a route needs specific treatments for addressing nonlinear dynamics.

The frequency-based formulation, also known as harmonic formulation, reads

$$\left(-\omega^2 \mathbf{M} + i\omega \mathbf{C} + \mathbf{K}\right) \mathcal{U} = \mathcal{F},$$

where $\mathcal{U}$ and $\mathcal{F}$ refer to the Fourier transform of the nodal displacement and force vectors, respectively $\mathbf{U}$ and $\mathbf{F}$.

Other than the difficulty to operate in nonlinear settings, an additional difficulty concerns the necessity of solving the problem for each angular frequency ω involved in the problem. This difficulty can be alleviated within the PGD rationale by considering the frequency as an extra-coordinate (a model parameter) as widely considered in our works. A more sophisticated approach was proposed in the context of the so-called VTCR [5, 6].

An added value is that the harmonic formulations avoids the necessity of constructing the basis $\mathbf{P}$ that requires the solution of the large-size eigenproblems.

Harmonic analysis allows to address parametric models in a very efficient way, from the use, for example, of separated representations where the parametric solution $\mathcal{U}(\omega, \boldsymbol{\mu})$ is expressed in the separated form

$$\mathcal{U}(\omega, \boldsymbol{\mu}) = \sum_i \mathcal{V}_i W_i(\omega) \prod_j M_i^j(\mu_j), \tag{31.1}$$

that can be constructed by using a purely algebraic procedure [7, 8].

Another advantage of the harmonic formulation is the possibility to consider more general dampings. It is important to note that even when considering complex nonlinear frequency dependent damping $\mathbf{C}(\omega)$, the problem in the frequency domain remains linear because here the frequency is a model parameter (or a model extra-coordinate) within the Proper Generalized Decomposition, PGD, framework (31.1).

Harmonic analysis, and its parametric version, were also successfully applied when addressing first order dynamics, as the one concerned by the heat equation [9].

31.4 Harmonic-Modal Hybrid Analysis

By considering the harmonic formulation with a proportional damping

$$\left(-\omega^2 \mathbf{M} + i\omega \mathbf{C} + \mathbf{K}\right) \mathcal{U} = \mathcal{F},$$

which when using the associated modal basis, with the normal modes grouped in matrix $\mathbf{P}$, using the transformation $\mathcal{U} = \mathbf{P}\boldsymbol{\xi}$ and taking into account the orthogonality conditions, reads

$$\left(-\omega^2 \mathbb{M} + i\omega \mathbb{C} + \mathbb{K}\right) \boldsymbol{\xi}(\omega) = \mathbf{P}^T \mathcal{F}(\omega) = \mathbf{f}(\omega)$$

leading to a system of N decoupled algebraic equations $\left(-\omega^2 + i\omega c_i + \kappa_i\right)\xi_i(\omega) = f_i(\omega)$, with $i = 1, 2, \ldots, N$, whose explicit solution reads

$$\xi_i(\omega) = \frac{f_i(\omega)}{-\omega^2 + i\omega c_i + \kappa_i}.$$

From the calculated $\boldsymbol{\xi}(\omega)$, the nodal amplitudes results $\mathcal{U}(\omega) = \mathbf{P}\boldsymbol{\xi}(\omega)$, which allows computing the nodal displacements $\mathbf{U}(t)$ by using the inverse Fourier transform [10, 11].

31.4.1 Nonlinear Models

Many engineering applications cannot be modeled as a linear system. In the nonlinear case, the general semi-discretized momentum balance can be written as

$$\mathbf{M}\frac{d^2\mathbf{U}(t)}{dt^2} + \mathbf{C}\frac{d\mathbf{U}(t)}{dt} + \mathbf{K}\mathbf{U}(t) - \mathbf{R}(\mathbf{U}) = \mathbf{F}(t), \qquad (31.2)$$

where $\mathbf{R}(\mathbf{U})$ is a nonlinear contribution that depends on $\mathbf{U}(t)$.

The simplest linearization considers the nonlinear term $\mathbf{R}(\mathbf{U})$ at the previous iteration of the nonlinear iteration, from which it results

$$\mathbf{M}\frac{d^2\mathbf{U}(t)}{dt^2} + \mathbf{C}\frac{d\mathbf{U}(t)}{dt} + \mathbf{K}\mathbf{U}(t) = \mathbf{R}(\mathbf{U}^-(t)) + \mathbf{F}(t).$$

Now, all the rationale previously introduced applies, with $\mathcal{F}(\omega)$ being now the Fourier transform of $\mathbf{R}(\mathbf{U}^-(t)) + \mathbf{F}(t)$.

Looking for enhancing the computational efficiency, the loading is expressed in the modal basis according to $\mathbf{f}(\omega) = \mathbf{P}^T \mathcal{F}(\omega)$, and then approximated according to

$$\mathbf{f}(\omega) = \sum_{l=1}^{L} \mathbf{f}(\omega_l) N_l(\omega),$$

with $N_l(\omega)$ the approximation functions in the frequency domain. Thus, the problem solution reads

$$\xi_i(\omega) = \frac{\sum_{l=1}^{L} f_i(\omega_l) N_l(\omega)}{-\omega^2 + i\omega c_i + \kappa_i}.$$

The expression above allows the offline calculation of the inverse transform of functions $N_l(\omega)$, making possible an almost real-time online reconstruction.

References

1. L. Boucinha, A. Gravouil, A. Ammar, Space-time proper generalized decompositions for the resolution of transient elastodynamic models. Comput. Methods Appl. Mech. Eng. **255**, 67–88 (2013)
2. L. Boucinha, A. Ammar, A. Gravouil, A. Nouy, Ideal minimal residual-based proper generalized decomposition for non-symmetric multi-field models—application to transient elastodynamics in space-time domain. Comput. Methods Appl. Mech. Eng. **273**, 56–76 (2014)
3. D. Gonzalez, E. Cueto, F. Chinesta, Real-time direct integration of reduced solid dynamics equations. Int. J. Numer. Methods Eng. **99**(9), 633–653 (2014)
4. T. Frahi, A. Falco, B. Vinh Mau, J.L. Duval, F. Chinesta, Empowering advanced parametric modes clustering from topological data analysis. Appl. Sci. **11**, 6554 (2021)
5. A. Barbarulo, P. Ladeveze, H. Riou, L. Kovalevsky, Proper generalized decomposition applied to linear acoustic: a new tool for broad band calculation. J. Sound Vib. **333**(11), 2422–2431 (2014)
6. A. Barbarulo, H. Riou, L. Kovalevsky, P. Ladeveze, PGD-VTCR: a reduced order model technique to solve medium frequency broad band problems on complex acoustical systems. J. Mech. Eng. **60**(5), 307–314 (2014)
7. M.H. Malik, D. Borzacchiello, F. Chinesta, P. Diez, Inclusion of frequency dependent parameters in power transmission lines simulation using harmonic analysis and proper generalized decomposition. Int. J. Numer. Model. Electron. Netw. Devices Fields **31**(5), e2331 (2018)
8. C. Germoso, J.L. Duval, F. Chinesta, Harmonic-modal hybrid reduced order model for the efficient integration of non-linear soil dynamics. Appl. Sci. **10**(19), 6778 (2020)
9. J.V. Aguado, A. Huerta, F. Chinesta, E. Cueto, Real-time monitoring of thermal processes by reduced order modelling. Int. J. Numer. Methods Eng. **102**(5), 991–1017 (2015)
10. M.H. Malik, D. Borzacchiello, J.V. Aguado, F. Chinesta, Advanced parametric space-frequency separated representations in structural dynamics: a harmonic-modal hybrid approach. CRAS Mecanique **346**(7), 590–602 (2018)
11. G. Quaranta, C. Argerich, R. Ibanez, J.L. Duval, E. Cueto, F. Chinesta, From linear to nonlinear PGD-based parametric structural dynamics. CRAS Mecanique **34**(5), 445–454 (2019)

Open Access This chapter is licensed under the terms of the Creative Commons Attribution-NonCommercial-NoDerivatives 4.0 International License (http://creativecommons.org/licenses/by-nc-nd/4.0/), which permits any noncommercial use, sharing, distribution and reproduction in any medium or format, as long as you give appropriate credit to the original author(s) and the source, provide a link to the Creative Commons license and indicate if you modified the licensed material. You do not have permission under this license to share adapted material derived from this chapter or parts of it.

The images or other third party material in this chapter are included in the chapter's Creative Commons license, unless indicated otherwise in a credit line to the material. If material is not included in the chapter's Creative Commons license and your intended use is not permitted by statutory regulation or exceeds the permitted use, you will need to obtain permission directly from the copyright holder.

Chapter 32
Space Separation

When addressing plate or shell geometries in which mathematical models are defined and must be solved, 3D approximations based on the use of meshes covering the domain are compromised, being one of its characteristic dimensions of the domain (the thickness in the case of plates and shell-like geometries) much smaller than the other characteristic dimensions, making difficult the use of good quality meshes.

When simple behaviors are considered (linear elasticity for instance), structural plate and shell theories were developed, and successfully applied. However, as soon as richer physics or geometries are concerned the validity of the different reduction hypotheses is compromised.

In these circumstances fully 3D seems compulsory. The in-plane-out-of-plane separated representations offer a valuable route able to compute the different unknown 3D fields without the necessity to introduce any hypothesis, and with a computational cost characteristic of standard 2D solutions.

32.1 In-Plane-Out-of-Plane Separated Representation of a Multilayered Plate

We assume a generic model, for the sake of simplicity the heat equation, defined in a plate domain $\Xi = \Omega \times \mathcal{I}$ with $\Omega \subset \mathbb{R}^2$ and $\mathcal{I} = [0, H] \subset \mathbb{R}$. Points $(x, y, z) \in \Xi$ are expressed as $(x, y, z) = (\mathbf{x}, z)$, with $\mathbf{x} = (x, y) \in \Omega$ and $z \in \mathcal{I}$.

Thus, the problem reads $\nabla \cdot (\mathbf{K} \nabla u) = 0$, whose weak form, assuming Dirichlet boundary conditions, reads

$$\int_{\Xi} \nabla u^* \cdot (\mathbf{K} \nabla u) \ d\Xi = 0,$$

with the test function u^* defined in an appropriate functional space.

The solution $u(x, y, z)$ is searched under the separated form

$$u(\mathbf{x}, z) \approx \sum_{j=1}^{N} X_j(\mathbf{x}) Z_j(z).$$

At the enrichment step $n - 1 < N$ the solution $u^{n-1}(\mathbf{x}, z)$ is already known, and at the present step n we look for the solution enrichment $X_n(\mathbf{x}) Z_n(z)$, such that

$$u^n(\mathbf{x}, z) = u^{n-1}(\mathbf{x}, z) + X_n(\mathbf{x}) Z_n(z),$$

with $u^*(\mathbf{x}, z) = X^*(\mathbf{x}) Z_n(z) + X_n(\mathbf{x}) \cdot Z^*(z)$.

With the above trial and test functions, the problem weak form reads

$$\int_\Xi \left(\begin{pmatrix} \tilde{\nabla} X^* \cdot Z_n \\ X^* \cdot \frac{dZ_n}{dz} \end{pmatrix} + \begin{pmatrix} \tilde{\nabla} X_n \cdot Z^* \\ X_n \cdot \frac{dZ^*}{dz} \end{pmatrix} \right) \cdot \left(\mathbf{K} \begin{pmatrix} \tilde{\nabla} X_n \cdot Z_n \\ X_n \cdot \frac{dZ_n}{dz} \end{pmatrix} \right) d\Xi$$

$$= - \int_\Xi \left(\begin{pmatrix} \tilde{\nabla} X^* \cdot Z_n \\ X^* \cdot \frac{dZ_n}{dz} \end{pmatrix} + \begin{pmatrix} \tilde{\nabla} X_n \cdot Z^* \\ X_n \cdot \frac{dZ^*}{dz} \end{pmatrix} \right) \cdot \mathbf{Q}^{n-1} \, d\Xi, \quad (32.1)$$

where $\tilde{\nabla}$ denotes the plane component of the gradient operator, i.e. $\tilde{\nabla} = \left(\frac{\partial}{\partial x}, \frac{\partial}{\partial y} \right)^T$ and $\mathbf{Q}^{n-1}$ denotes the flux at erichment $n - 1$

$$\mathbf{Q}^{n-1} = \mathbf{K} \sum_{j=1}^{n-1} \begin{pmatrix} \tilde{\nabla} X_j(\mathbf{x}) \cdot Z_j(z) \\ X_j(\mathbf{x}) \cdot \frac{dZ_j(z)}{dz} \end{pmatrix}.$$

Now, to search the couple of functions $X_n(\mathbf{x})$ and $Z_n(z)$ the alternated direction fixed point algorithm (already described) is used.

Such space separated representation was successfully applied in many applicative domains. Thus, 3D elasticity defined in plates, shell and beams was addressed in [1–4]. Space separation in elastodynamics was considered [5], in thermal models in [6, 7], in flow problems in [8–12] and in electromagnetic problems in [13]. A minimally-intrusive procedure was proposed [14, 15] for structural mechanics applications.

32.2 Addressing Non-separable Domains: Smart Mapping

To address geometries not directly separable, the use of the so-called smart mapping was proposed in some of our former works, summarized here.

For the sake of simplicity, the solution of a two-dimensional heat transfer problem is considered. The problem is defined in a thin domain $\Omega = (0, L) \times (0, H)$,

32.2 Addressing Non-separable Domains: Smart Mapping

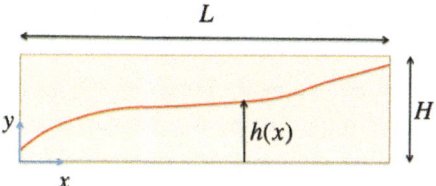

Fig. 32.1 Domain containing a nonplanar internal boundary

with $H \ll L$, where a non-planar internal boundary expressed by the function $h(x)$ (sketched in Fig. 32.1) separates the upper and bottom domains Ω_u and Ω_b, with thermal conductivities K_u and K_b respectively.

The temperature field $u(x, y)$ is he solution of the heat transfer problem defined in Ω, whose weak form reads

$$\int_\Omega K(\mathbf{x}) \nabla u^* \cdot \nabla u \, d\mathbf{x} = 0,$$

where the temperature is assumed enforced in the top and bottom boundaries, and with the heat flux vanishing on the lateral ones.

To address the domain degeneracy, a space separated representation is employed:

$$u(x, y) \approx \sum_{i=1}^{N} X_i(x) Y_i(y),$$

however, that separated representation is compromised by the presence of the non-planar interface, that implies a hardly separable conductivity field.

Conductivity can always be separated by invoking, for instance, the singular value decomposition (SVD)

$$K(x, y) \approx \sum_{k=1}^{K} F_k(x) G_k(y),$$

however, the number of modes K increases too much, and with it the operators involved in the weak form.

In order to circumvent this issue, domains Ω_u and Ω_b are mapped into, respectively, $\mathcal{R}_u$ and $\mathcal{R}_b$, according to:

1. *Mapping Ω_b into $\mathcal{R}_b$.*
   First, $\mathbf{r} = (r, s) \in \mathcal{R}_b = (0, L) \times (0, 1)$ reads

$$\begin{cases} x = r, \\ y = s\, h(r), \end{cases}$$

or, its inverse,

$$\begin{cases} r = x, \\ s = \frac{y}{h(x)}. \end{cases}$$

both fully separable, and with them, the Jacobian components.

2. Mapping Ω_u into $\mathcal{R}_u$.

Equivalently, for Ω_u

$$\begin{cases} x = r, \\ y = (s-1)(H - h(r)) + h(r), \end{cases}$$

with $(r, s) \in \mathcal{R}_b = (0, L) \times (1, 2)$.

The problem can be solved from a sequence of one-dimensional problems in the r and s coordinate domains, by using the standard PGD procedure. Adapted mappings were deeply considered in some of our former works, among them [12, 16–18].

32.3 NURBS-Based PGD

32.3.1 NURBS-Based Geometry Description

NURBS enable an exact representation of the geometry for most curves, surfaces and volumes encountered in engineering applications, being nowadays a usual technology employed in CAD.

NURBS result from weighted combinations of B-spline functions. To define a set of n B-spline functions of order p in a univariate parametric space $\xi \in [0, 1]$, the knot vector κ_ξ is defined as follows

$$\kappa_\xi = [\xi_1, \xi_2, \ldots, \xi_{n+p+1}]^T.$$

With $N_{ap}(\xi)$ the B-spline basis function of order p in the a-th knot span $\xi \in [\xi_a, \xi_{a+1})$, the following recursive equations are employed to compute the univariate B-spline basis function $N_{ap}(\xi)$

$$N_{a0}(\xi) = \begin{cases} 1 & \xi_a \le \xi < \xi_{a+1} \\ 0 & \text{otherwise} \end{cases},$$

for $p = 0$, and for $p > 0$

$$N_{ap}(\xi) = \frac{\xi - \xi_a}{\xi_{a+p} - \xi_a} N_{a(p-1)} + \frac{\xi_{a+p+1} - \xi}{\xi_{a+p+1} - \xi_{a+1}} N_{(a+1)(p-1)}$$

32.3 NURBS-Based PGD

A proper choice of the knot vectors allows rich approximations and enough flexibility to describe complex geometries.

The NURBS basis functions can be obtained using a rational weighted sum of the B-splines basis functions. With w_i the weight, the univariate NURBS basis function, $R_i^p(\xi)$ reads

$$R_i^p(\xi) = \frac{N_{ip}(\xi) w_i}{\sum_{\alpha=1}^n N_{\alpha p}(\xi) w_\alpha}.$$

The bivariate ou trivariate NURBS, $R_{ij}^{pq}(\xi, \eta)$ and $R_{ijk}^{pqr}(\xi, \eta, \zeta)$ are obtained by tensor products

$$R_{ij}^{pq}(\boldsymbol{\xi}) = \frac{N_{ip}(\xi) N_{jq}(\eta) w_{ij}}{\sum_{\alpha=1}^n \sum_{\beta=1}^m N_{\alpha p}(\xi) N_{\beta q}(\eta) w_{\alpha\beta}},$$

and

$$R_{ijk}^{pqr}(\boldsymbol{\xi}) = \frac{N_{ip}(\xi) N_{jq}(\eta) N_{kr}(\zeta) w_{ijk}}{\sum_{\alpha=1}^n \sum_{\beta=1}^m \sum_{\gamma=1}^l N_{\alpha p}(\xi) N_{\beta q}(\eta) N_{\gamma r}(\zeta) w_{\alpha\beta\gamma}},$$

with p, q, and r the order of the B-splines in directions ξ, η, and ζ and $\boldsymbol{\xi} = (\xi, \eta, \zeta)$ the coordinates vector in the computational domain.

Using the NURBS basis functions just described, a curve C, surface S and volume V can be expressed from

$$\boldsymbol{x} = \sum_{i=1}^n R_i^p(\boldsymbol{\xi}) \boldsymbol{P}_i \qquad \boldsymbol{x} \in C,$$

$$\boldsymbol{x} = \sum_{i=1}^n \sum_{j=1}^m R_{ij}^{pq}(\boldsymbol{\xi}) \boldsymbol{P}_{ij} \qquad \boldsymbol{x} \in S,$$

and

$$\boldsymbol{x} = \sum_{i=1}^n \sum_{j=1}^m \sum_{k=1}^l R_{ijk}^{pqr}(\boldsymbol{\xi}) \boldsymbol{P}_{ijk} \qquad \boldsymbol{x} \in V,$$

respectively, where, $\boldsymbol{P}$ refers to the control points (the vertices of the so-called control net) in the physical domain.

These expressions represent the application that maps any point $\boldsymbol{\xi}$ in the computational domain Ω_ξ to a point $\mathbf{x}$ in the physical domain $\Omega_\mathbf{x}$.

Thus, curves could be easily transformed into a unit segment, surfaces into a unit square and volumes into a unit cube by carefully choosing the knots vector in each direction, the degree and number and position of the control points.

These unit segment, square or cube are referred as computational domains and are specially appealing in the PGD framework because they enable the use of the

PGD-based space separation, that solves two and three dimensional problems as a sequence of one-dimensional problems, with the consequent computational benefits.

32.4 Non-intrusive Space Separation

Until now, fields were expressed using a separated representation and then the unknown functions involved in it computed. Here, in order to reduce intrusiveness, we proceed in the opposite manner, we discretize the problem first, by using any well experienced discretization technique, and then, the discrete form of the unknown field is expressed in a separated form, facilitating the numerical solution of the linear system [19, 20].

When considering a standard 3D finite element approximation and discretization, the resulting discrete form reads $\mathbf{GU} = \mathbf{F}$, where vector $\mathbf{U}$ contains the nodal variables, assumed, for the sake of simplicity, scalar (e.g. temperature).

A layered mesh along the domain thickness is employed, composed of $\mathcal{T}$ layers, with $\mathbf{U}_i$, $i = 1, \ldots, \mathcal{T}$ referring to the nodal values associated to nodes belonging to the layer $\mathcal{L}_i$, $i = 1, \ldots, \mathcal{T}$. Thus, by employing an adequate nodal re-numbering, the discrete system can be rewritten as

$$\begin{pmatrix} \mathbf{G}_{11} & \mathbf{G}_{12} & \cdots & \mathbf{G}_{1\mathcal{T}} \\ \mathbf{G}_{21} & \mathbf{G}_{22} & \cdots & \mathbf{G}_{2\mathcal{T}} \\ \vdots & \vdots & \ddots & \vdots \\ \mathbf{G}_{\mathcal{T}1} & \mathbf{G}_{\mathcal{T}2} & \cdots & \mathbf{G}_{\mathcal{T}\mathcal{T}} \end{pmatrix} \begin{pmatrix} \mathbf{U}_1 \\ \mathbf{U}_2 \\ \vdots \\ \mathbf{U}_\mathcal{T} \end{pmatrix} = \begin{pmatrix} \mathbf{F}_1 \\ \mathbf{F}_2 \\ \vdots \\ \mathbf{F}_\mathcal{T} \end{pmatrix}.$$

Now, inspired by the in-plane-out-of-plane separated representation, we consider

$$\mathbf{U} = \sum_{i=1}^{N} \mathbf{V}_i \otimes \mathbf{W}_i,$$

where the size of vectors $\mathbf{V}_i$ is the number of in-plane degrees of freedom whereas the size of vectors $\mathbf{W}_i$ is the number of layers, i.e. $\mathcal{T}$. The involved functions are obtained by employing the usual separated representation constructor.

The just described procedure deserves the following important comments:

- Matrix $\mathbf{G}$ is agnostic with respect to the employed discretization technique. In particular it can be obtained from the use of any finite element software using any type of finite element. In this sense the just described procedure becomes much less intrusive than the one resulting of introducing the separated representation before its discretization (usual PGD);
- In the case of nonlinear models matrix $\mathbf{G}$ corresponds to the tangent matrix and again can be computed by an external discretization software;

- The proposed strategy could be viewed as an iterative linear solver that constructs a separated representation of the nodal solution;
- The solution procedure only involves matrix products and linear system solutions (of size and complexity characteristic of the 2D and 1D problems involved in the decomposition) that can be efficiently performed using massively parallel computing architectures;
- The solution procedure can be viewed as a domain decomposition technique, where continuity is ensured from the approximation, for instance the finite element, employed in the construction of matrix **G**, and in which the information spreads all along the whole domain at each iteration.
- This procedure can be applied to any partition of the vector containing the nodal unknowns as soon as all them contain the same number of nodes.

References

1. B. Bognet, A. Leygue, F. Chinesta, A. Poitou, F. Bordeu, Advanced simulation of models defined in plate geometries: 3D solutions with 2D computational complexity. Comput. Methods Appl. Mech. Eng. **201**, 1–12 (2012)
2. E. Giner, B. Bognet, J.J. Rodenas, A. Leygue, J. Fuenmayor, F. Chinesta, The Proper Generalized Decomposition (PGD) as a numerical procedure to solve 3D cracked plates in linear elastic fracture mechanics. Int. J. Solid Struct. **50**(10), 1710–1720 (2013)
3. B. Bognet, A. Leygue, F. Chinesta, Separated representations of 3D elastic solutions in shell geometries. Adv. Model. Simul. Eng. Sci. **1**, 4 (2014)
4. F. Bordeu, Ch. Ghnatios, D. Boulze, B. Carles, D. Sireude, A. Leygue, F. Chinesta, Parametric 3D elastic solutions of beams involved in frame structures. Adv. Aircr. Spacecr. Sci. **2**(3), 233–248 (2015)
5. G. Quaranta, B. Bognet, R. Ibanez, A. Tramecon, E. Haug, F. Chinesta, A new hybrid explicit/implicit in-plane-out-of-plane separated representation for the solution of dynamic problems defined in plate-like domains. Comput. Struct. **210**, 135–144 (2018)
6. A. Leygue, F. Chinesta, M. Beringhier, T.L. Nguyen, J.C. Grandidier, F. Pasavento, B. Schrefler, Towards a framework for non-linear thermal models in shell domains. Int. J. Numer. Meth. Heat Fluid Flow **23**(1), 55–73 (2013)
7. F. Chinesta, A. Leygue, B. Bognet, Ch. Ghnatios, F. Poulhaon, F. Bordeu, A. Barasinski, A. Poitou, S. Chatel, S. Maison-Le-Poec, First steps towards an advanced simulation of composites manufacturing by automated tape placement. Int.J. Mater. Form. **7**(1), 81–92 (2014)
8. R. Ibanez, E. Abisset-Chavanne, F. Chinesta, A. Huerta, 3D mixed formulation for simulating squeeze flows in multiaxial laminates. Int. J. Mater. Form. **10**(5), 653–669 (2017)
9. D. Canales, A. Leygue, F. Chinesta, I. Alfaro, D. Gonzalez, E. Cueto, E. Feulvarch, J.M. Bergheau, In-plane-out-of-plane separated representations of updated-Lagrangian descriptions of thermomechanical models defined in plate domains. CRAS **344**(4–5), 225–235 (2016)
10. Ch. Ghnatios, F. Chinesta, Ch. Binetruy, The squeeze flow of composite laminates. Int. J. Mater. Form. **8**, 73–83 (2015)
11. Ch. Ghnatios, E. Abisset-Chavanne, Ch. Binetruy, F. Chinesta, S. Advani, 3D modeling of squeeze flow of multiaxial laminates. J. Nonnewton. Fluid Mech. **234**, 188–200 (2016)
12. A. Ammar, Ch. Ghnatios, F. Delplace, A. Barasinski, J.L. Duval, E. Cueto, F. Chinesta, On the effective conductivity and the apparent viscosity of a thin-rough polymer interface using PGD-based separated representations. Int. J. Numer. Meth. Eng. **121**(23), 5256–5274 (2020)
13. H. Tertrais, R. Ibanez, A. Barasinski, Ch. Ghnatios, F. Chinesta, On the proper generalized decomposition applied to microwave processes involving multilayered components. Math. Comput. Simul. **156**, 347–363 (2019)

14. G. Quaranta, M. Ziane, F. Daim, E. Abisset-Chavanne, J.L. Duval, F. Chinesta, On the coupling of local 3D solutions and global 2D shell theory in structural mechanics. AMSES **6**(1) (2019)
15. G. Quaranta, M. Ziane, E. Haug, J.L. Duval, F. Chinesta, A minimally-intrusive fully 3D separated plate formulation in computational structural mechanics. Adv. Model. Simul. Eng. Sci. **6**(11) (2019)
16. Ch. Ghnatios, G. Xu, M. Visonneau, A. Leygue, F. Chinesta, A. Cimetiere, On the space separated representation when addressing the solution of PDE in complex domains. Discret. Contin. Dynam. Syst. **9**(2), 475–500 (2016)
17. C. Ghnatios, E. Abisset-Chavanne, A. Ammar, E. Cueto, J.L. Duval, F. Chinesta, Advanced spatial separated representations. Comput. Methods Appl. Mech. Eng. **354**, 802–819 (2019)
18. C. Ghnatios, E. Cueto, A. Falco, J.L. Duval, F. Chinesta, Spurious-free interpolations for non-intrusive PGD-based parametric solutions: application to composites forming processes. Int. J. Mater. Form. (2020)
19. A. Leon, S. Mueller, P. de Luca, R. Said, J.L. Duval, F. Chinesta, Non-intrusive proper generalized decomposition involving space and parameters: application to the mechanical modeling of 3D woven fabrics. Adv. Model. Simul. Eng. Sci. **6**(13) (2019)
20. C. Germoso, G. Quaranta, J.L. Duval, F. Chinesta, Non-intrusive in-plane-out-of-plane separated representation in 3D parametric elastodynamics. Computation **8**(3), 78 (2020)

Open Access This chapter is licensed under the terms of the Creative Commons Attribution-NonCommercial-NoDerivatives 4.0 International License (http://creativecommons.org/licenses/by-nc-nd/4.0/), which permits any noncommercial use, sharing, distribution and reproduction in any medium or format, as long as you give appropriate credit to the original author(s) and the source, provide a link to the Creative Commons license and indicate if you modified the licensed material. You do not have permission under this license to share adapted material derived from this chapter or parts of it.

The images or other third party material in this chapter are included in the chapter's Creative Commons license, unless indicated otherwise in a credit line to the material. If material is not included in the chapter's Creative Commons license and your intended use is not permitted by statutory regulation or exceeds the permitted use, you will need to obtain permission directly from the copyright holder.

Part IV
Around Data Assimilation and Twining

Chapter 33
Data Assimilation, Inverse Analysis and Control

Dynamic Data-Driven Application Systems, DDDAS, represents a new paradigm in the field of applied sciences and engineering, and in particular in Simulation-Based Engineering Sciences, SBES. By DDDAS we mean a set of techniques that allow to link simulation tools with measurement devices for real-time control of systems and processes.

A DDDAS includes different constituent blocks: (i) A set of (possibly) heterogeneous simulation models; (ii) A system to handle data obtained from both static and dynamic sources; (iii) Algorithms to efficiently predict system behavior by solving the models; (iv) Software infrastructure to integrate the data, model predictions, control algorithms, etc.

This section revisits some technologies involved in DDDAS.

33.1 Optimal Control

In the context of observability and controllability of dynamical systems involved in digital systems, optimal control is a major protagonist. This section revisits its foundations.

First, we assume a generic dynamical system:

$$\dot{\mathbf{x}}(t) = \mathbf{f}(\mathbf{x}(t), \mathbf{u}(t), t), \quad t \in (0, t_f]$$

with $\mathbf{x} \in \mathbb{R}^N$, $\mathbf{u} \in \mathbb{R}^R$ and $\mathbf{x}(t=0) = \mathbf{x}_0$.

The quantity to be minimized, $J(\mathbf{x}(t), \mathbf{u}(t), \mathbf{x}_0)$, is assumed composed of two contributions: (i) a terminal cost $\mathcal{K}(\mathbf{x}(t_f), t_f)$; and (ii) the instantaneous cost contribution $\mathcal{L}(\mathbf{x}(t), \mathbf{u}(t), t)$:

$$J(\mathbf{x}(t), \mathbf{u}(t), \mathbf{x}_0) = \mathcal{K}(\mathbf{x}(t_f), t_f) + \int_0^{t_f} \mathcal{L}(\mathbf{x}(t), \mathbf{u}(t), t)\, dt.$$

Under certain regularity conditions, the fulfillment of the dynamical system can be added to the cost function from the use of a Lagrange multiplier, i.e.,

$$\tilde{J}(\mathbf{x}(t), \mathbf{u}(t), \mathbf{x}_0)$$
$$= \mathcal{K}(\mathbf{x}(t_f), t_f) + \int_0^{t_f} \left[\mathcal{L}(\mathbf{x}(t), \mathbf{u}(t), t) + \boldsymbol{\lambda}^T \left(\mathbf{f}(\mathbf{x}(t), \mathbf{u}(t), t) - \dot{\mathbf{x}}(t) \right) \right] dt$$
$$= \mathcal{K}(\mathbf{x}(t_f), t_f) + \int_0^{t_f} \left[\mathcal{H}(\mathbf{x}(t), \mathbf{u}(t), \boldsymbol{\lambda}(t), t) - \boldsymbol{\lambda}^T(t) \dot{\mathbf{x}}(t) \right] dt,$$

with $\mathcal{H}(\mathbf{x}(t), \mathbf{u}(t), \boldsymbol{\lambda}(t), t) \equiv \mathcal{L}(\mathbf{x}(t), \mathbf{u}(t), t) + \boldsymbol{\lambda}^T \mathbf{f}(\mathbf{x}(t), \mathbf{u}(t), t)$.

The term involving the state time derivative $\dot{\mathbf{x}}$ can be integrated by parts

$$-\int_{t_0}^{t_f} \boldsymbol{\lambda}^T(t) \dot{\mathbf{x}}(t) \, dt = -\boldsymbol{\lambda}^T(t_f) \mathbf{x}(t_f) + \boldsymbol{\lambda}^T(t_0) \mathbf{x}(t_0) + \int_{t_0}^{t_f} \dot{\boldsymbol{\lambda}}^T \mathbf{x}(t) \, dt,$$

from which the extended cost function can be rewritten in the form

$$\tilde{J}(\mathbf{x}(t), \mathbf{u}(t), \mathbf{x}_0)$$
$$= \{ \mathcal{K}(\mathbf{x}(t_f), t_f) - \boldsymbol{\lambda}^T(t_f) \mathbf{x}(t_f) \} + \boldsymbol{\lambda}^T(t_0) \mathbf{x}_0$$
$$+ \int_0^{t_f} \{ \mathcal{H}(\mathbf{x}(t), \mathbf{u}(t), \boldsymbol{\lambda}(t), t) + \dot{\boldsymbol{\lambda}}^T(t) \mathbf{x}(t) \} \, dt.$$

Considering that $\mathbf{u}^*(t)$ is an optimal control, the associated trajectory is denoted by $\mathbf{x}^*(t)$. Thus considering

$$\begin{cases} \mathbf{u}(t) = \mathbf{u}^*(t) + \delta\mathbf{u}(t) \\ \mathbf{x}(t) = \mathbf{x}^*(t) + \delta\mathbf{x}(t) \end{cases},$$

the following inequality is obtained

$$\tilde{J}(\mathbf{x}^*(t), \mathbf{u}^*(t), \mathbf{x}_0) \leq \tilde{J}(\mathbf{x}^*(t) + \delta\mathbf{x}(t), \mathbf{u}^*(t) + \delta\mathbf{u}(t), \mathbf{x}_0), \quad \forall \delta\mathbf{x}(t), \forall \delta\mathbf{u}(t).$$

The optimality derives from:

$$\tilde{J}(\mathbf{x}^*(t) + \delta\mathbf{x}(t), \mathbf{u}^*(t) + \delta\mathbf{u}(t), \mathbf{x}_0) - \tilde{J}(\mathbf{x}^*(t), \mathbf{u}^*(t), \mathbf{x}_0)$$
$$= \int_{t_0}^{t_f} \left\{ \left(\frac{\partial \mathcal{H}}{\partial \mathbf{x}} \right)^T (\mathbf{x}^*, \mathbf{u}^*, t, \boldsymbol{\lambda}(t)) \delta\mathbf{x} + \left(\frac{\partial \mathcal{H}}{\partial \mathbf{u}} \right)^T (\mathbf{x}^*, \mathbf{u}^*, t, \boldsymbol{\lambda}(t)) \delta\mathbf{u} + \dot{\boldsymbol{\lambda}}^T \delta\mathbf{x} \right\} dt$$
$$+ \left[\left(\frac{\partial \mathcal{K}}{\partial \mathbf{x}} \right)^T (\mathbf{x}^*(t_f), t_f) - \boldsymbol{\lambda}^T(t_f) \right] \delta\mathbf{x}(t_f) + \boldsymbol{\lambda}^T(t_0) \delta\mathbf{x}_0 \geq 0, \quad (33.1)$$

and reads:

- Since the initial state is fixed $\delta \mathbf{x}_0 = \mathbf{0}$, no condition on $\lambda(t_0)$ is required.
- The final state being free, i.e., $\delta \mathbf{x}(t_f) \neq \mathbf{0}$, then, the equality

$$\lambda(t_f) = \frac{\partial \mathcal{K}}{\partial \mathbf{x}}(\mathbf{x}^*(t_f), t_f),$$

becomes compulsory.
- Since the inequality Eq. (33.1) remains valid for any $\delta \mathbf{x}$, the Lagrange multiplier must verify

$$\dot{\lambda}(t) = -\frac{\partial \mathcal{H}}{\partial \mathbf{x}}(\mathbf{x}^*, \mathbf{u}^*, t, \lambda(t)).$$

- Finally, the arbitrariness with respect to $\delta \mathbf{u}$ leads to:

$$\frac{\partial \mathcal{H}}{\partial \mathbf{u}}(\mathbf{x}^*, \mathbf{u}^*, t, \lambda(t)) = \mathbf{0}.$$

33.2 Assimilation by Tikhonov Regularization of PGD-Based Parametric Solutions

In what follows, for the sake of simplicity and without loss of generality we assume a single parameter μ concerned by the parametric model, whose solution is available in a separated form (e.g. obtained within the PGD setting).

Thus, the solution of the parametric problem expresses

$$u^{\text{PGD}}(\mathbf{x}, \mu) = \sum_{i=1}^{N} X_i(\mathbf{x}) M_i(\mu).$$

We also assume that a series of measures are available at the sensors location $\mathbf{s}_i$, $i = 1, \ldots, M$, $u^{\text{exp}}(\mathbf{s}_i)$. We would like to assimilate the data in order to infer a solution such that it represents the measures provided by the sensors, $u^{\text{exp}}(\mathbf{s}_i)$. The number and optimal location of the sensors will be addressed later.

In what follows we consider two routes [1]: (i) one that considers directly the PGD-based parametric solution; and (ii) one making use of a reduced basis extracted from the parametric PGD solution.

33.2.1 Assimilation Based on the Complete PGD Solution

In this case it is assumed that the assimilated solution at the sensor locations, reads $u^{\text{ass}}(\mathbf{s}_i) = u^{\text{PGD}}(\mathbf{s}_i, \mu^{\text{ass}})$.

The parameter μ^{ass} is identified by minimizing the functional

$$\mathcal{J}(\mu) = \sum_{i=1}^{M} \left(u^{\text{exp}}(\mathbf{s}_i) - u^{\text{PGD}}(\mathbf{s}_i, \mu) \right)^2.$$

The main issues related to such a procedure are:

- The minimization of the functional leads to a nonlinear equation that should be solved, for control purposes, in almost real time.
- It was assumed that both the mathematical model and its PGD solution accurately describe the physical phenomena. In some practical situations this assumption could be strong since the physical phenomena could not be perfectly represented by the mathematical model. Moreover, measures can also contains an amount of error.
- In complex situations, the solution is extremely sensitive to the value of the parameter.

33.2.2 Assimilation Based on a PGD-Based Reduced Basis

We select a number $R < N$ of space functions involved in the PGD-solution: $\{X_1(\mathbf{x}), \ldots, X_R(\mathbf{x})\}$; that allows defining the new functional

$$\mathcal{J}'(\boldsymbol{\alpha}) = \sum_{i=1}^{M} \left(u^{\text{exp}}(\mathbf{s}_i) - \sum_{j=1}^{R} X_j(\mathbf{s}_i) \alpha_j \right)^2.$$

The α coefficients allow the best fit in a least square sense. Since the PGD solution is available, the functional is modified by introducing the parametric functions according to:

$$\mathcal{J}''(\mu, \boldsymbol{\epsilon}) = \sum_{i=1}^{M} \left(u^{\text{exp}}(\mathbf{s}_i) - \sum_{j=1}^{R} X_j(\mathbf{s}_i)(M_j(\mu) + \epsilon_j) \right)^2,$$

where ϵ_i also takes into account the possible discrepancies with the model.

Additionally, a weighted regularization term is added to the functional

$$\mathcal{J}^{\text{reg}}(\mu, \boldsymbol{\epsilon}) = \sum_{i=1}^{M} \left(u^{\text{exp}}(\mathbf{s}_i) - \sum_{j=1}^{R} X_j(\mathbf{s}_i)(M_j(\mu) + \epsilon_j) \right)^2 + \tau \sum_{j=1}^{R} \epsilon_j^2.$$

The value of τ plays an important role and depends on the application. Large values of τ provide an assimilated solution closer to the solution provided by the PGD. On the contrary, small values of τ penalize the discrepancy with the measurements. Therefore, the choice of τ depends on the degree of confidence associated with the PGD solution or with the measures.

Introducing $\tau > 0$ in the formulation, allows even considering R > M. This is not possible without the regularization term. In order to obtain the values of μ^{ass} and ϵ, functional $\mathcal{J}^{\text{reg}}(\mu, \epsilon)$ is enforced to be stationary.

33.3 Optimal Sensor Placement for PGD-Based Data Assimilation

We first consider the optimal placement of one (the first) sensor. If we restrict the PGD solution to the first mode it results

$$u^{\text{exp}} = X_1(\mathbf{x}) M_1(\mu^{\text{ass}}) + \delta,$$

being δ the discrepancy. This expression can be rewritten as

$$M_1(\mu^{\text{ass}}) = \frac{u^{\text{exp}} - \delta}{X_1(\mathbf{x})},$$

that in order to identify the parameter μ^{ass} while minimizing the effect of the just introduced discrepancy δ, recommends locating the sensor where $X_1(\mathbf{x})$ is maximum, i.e.

$$\mathbf{s}_1 = \text{argmax}_{\mathbf{x}} |X_1(\mathbf{x})|.$$

Then, by defining the residual $R_2(\mathbf{x}) = X_2(\mathbf{x}) - d X_1(\mathbf{x})$, with d satisfying $R_2(\mathbf{s}_1) = 0$, the second sensor should be placed where $R_2(\mathbf{x})$ is maximum. Thus, clearly appears that the sensors optimal location coincides with the *magic points* related to the PGD-based reduced basis, according to the DEIM rationale.

33.4 Bayesian Inverse Analysis and Data-Assimilation

From a parametric solution $u(\mathbf{x}, \boldsymbol{mu})$, and some collected data at certain locations $u^{\text{exp}}(\mathbf{x}_i), i = 1, \ldots, M$, one is interested in inferring the associated parameters value. In what follows we assume a single parameter μ involved in the physical model, and in its parametric solution $u(\mathbf{x}, \mu)$. In this context Bayesian techniques are very valuable for data assimilation [1, 2].

33.4.1 PGD-Likelihood Maximization in Absence of Priors

In absence of priors, that is, when nothing is known on the probability distribution of μ, as discussed in Chap. 25, the choice of the sampling is crucial, in particular when the likelihood is too flat.

The identification procedure proceeds as follows: from the first collected data at point $\mathbf{x}_1$, $u^{\text{exp}}(\mathbf{x}_1)$, one is tempted to find the value of the parameter μ^* that ensures $u(\mathbf{x}_1, \mu^*) = u^{\text{exp}}(\mathbf{x}_1)$. When many data are available $u^{\text{exp}}(\mathbf{x}_i)$, $i = 1, \ldots, M$, the best parameter choice μ^* results

$$\mu^* = \arg\min_{\mu} \| \sum_{i=1}^{M} (u(\mathbf{x}_i, \mu) - u^{\text{exp}}(\mathbf{x}_i)) \|_p,$$

where different norms can be considered. The L1-norm ($p = 1$), is more robust in presence of outliers.

33.4.2 Accounting for Priors in Bayesian Settings

In presence of priors, that is, existing knowledge on the parameter probability distribution, the identification process proceeds as discussed in Chap. 12.

33.5 Data-Assimilation Based on Kalman Filters

Kalman filters are widely used in data prediction and data-assimilation problems, when the state evolution and the measures contain a noticeable noise. In this section we try to summarize the main steps of its derivation.

We assume the dynamics expressible from

$$\mathbf{x}_{k+1} = \mathbf{\Phi} \mathbf{x}_k + \omega_k,$$

with index $\bullet_k$ referring to the considered variable at time $t_k \equiv k\Delta t$, $\mathbf{x} \in \mathbb{R}^N$ and the noise $\omega \in \mathbb{R}^N$.

Observation of the state is modeled from

$$\mathbf{z}_k = \mathbf{H} \mathbf{x}_k + \mathbf{v}_k, \qquad (33.2)$$

with $\mathbf{z} \in \mathbb{R}^M$ and its associated observation noise $\mathbf{v} \in \mathbb{R}^M$.

The covariances of both noises, assumed to be stationary, are expressed from:

$$\begin{cases} \mathbf{Q} = \mathbb{E}[\omega_k \omega_k^T] \\ \mathbf{R} = \mathbb{E}[\mathbf{v}_k \mathbf{v}_k^T] \end{cases}.$$

If we define the estimation of $\mathbf{x}_k$ as $\hat{\mathbf{x}}_k$, the error at time t_k is defined from $\mathbf{e}_k = \mathbf{x}_k - \hat{\mathbf{x}}_k$, and the error covariance at time t_k is given by

$$\mathbf{P}_k = \mathbb{E}[\mathbf{e}_k \mathbf{e}_k^T]. \qquad (33.3)$$

33.5 Data-Assimilation Based on Kalman Filters

By notting by $\hat{\mathbf{x}}'_k$ the prior estimate of $\hat{\mathbf{x}}_k$, the Kalman gain at time t_k, $\mathbf{K}_k$ is defined from

$$\hat{\mathbf{x}}_k = \hat{\mathbf{x}}'_k + \mathbf{K}_k(\mathbf{z}_k - \mathbf{H}\hat{\mathbf{x}}'_k). \tag{33.4}$$

Replacing Eq. (33.2) into Eq. (33.4), it results:

$$\hat{\mathbf{x}}_k = \hat{\mathbf{x}}'_k + \mathbf{K}_k(\mathbf{H}\mathbf{x}_k + \mathbf{v}_k - \mathbf{H}\hat{\mathbf{x}}'_k),)$$

that introduced in the expression of the error covariance, Eq. (33.3), reads

$$\mathbf{P}_k = \mathbb{E}[[(\mathbf{I} - \mathbf{K}_k\mathbf{H})(\mathbf{x}_k - \hat{\mathbf{x}}'_k) - \mathbf{K}_k\mathbf{v}_k][[(\mathbf{I} - \mathbf{K}_k\mathbf{H})(\mathbf{x}_k - \hat{\mathbf{x}}'_k) - \mathbf{K}_k\mathbf{v}_k]^T].$$

Using the fact that the error of the prior estimate and the one of the measurements are uncorrelated, the previous equation can be rewritten as

$$\mathbf{P}_k = (\mathbf{I} - \mathbf{K}_k\mathbf{H})\mathbb{E}[(\mathbf{x}_k - \hat{\mathbf{x}}'_k)(\mathbf{x}_k - \hat{\mathbf{x}}'_k)^T](\mathbf{I} - \mathbf{K}_k\mathbf{H})^T + \mathbf{K}_k\mathbb{E}[\mathbf{v}_k\mathbf{v}_k^T]\mathbf{K}_k^T,$$

or using the prior error covariance $\mathbf{P}'_k$

$$\mathbf{P}_k = (\mathbf{I} - \mathbf{K}_k\mathbf{H})\mathbf{P}'_k(\mathbf{I} - \mathbf{K}_k\mathbf{H})^T + \mathbf{K}_k\mathbb{E}[\mathbf{v}_k\mathbf{v}_k^T]\mathbf{K}_k^T =$$

$$(\mathbf{I} - \mathbf{K}_k\mathbf{H})\mathbf{P}'_k(\mathbf{I} - \mathbf{K}_k\mathbf{H})^T + \mathbf{K}_k\mathbf{R}\mathbf{K}_k^T =$$

$$\mathbf{P}'_k - \mathbf{K}_k\mathbf{H}\mathbf{P}'_k - \mathbf{P}'_k\mathbf{H}^T\mathbf{K}_k^T + \mathbf{K}_k(\mathbf{H}\mathbf{P}'_k\mathbf{H}^T + \mathbf{R})\mathbf{K}_k^T. \tag{33.5}$$

The trace, $\text{Tr}(\cdot)$, of that covariance matrix represents the sum of the mean squared errors, and reads

$$\text{Tr}[\mathbf{P}^k] = \text{Tr}[\mathbf{P}'_k] - 2\text{Tr}[\mathbf{K}_k\mathbf{H}\mathbf{P}'_k] + \text{Tr}[\mathbf{K}_k(\mathbf{H}\mathbf{P}'_k\mathbf{H}^T + \mathbf{R})\mathbf{K}_k^T],$$

where the fact that the trace of a matrix is equal to the one of its transpose was considered.

The trace (sum of the mean squared errors) minimization reads:

$$\frac{d\text{Tr}[\mathbf{P}^k]}{d\mathbf{K}_k} = -2(\mathbf{H}\mathbf{P}'_k)^T + 2\mathbf{K}_k(\mathbf{H}\mathbf{P}'_k\mathbf{H}^T + \mathbf{R}) = 0. \tag{33.6}$$

Remark. The derivative of the trace of a product of two matrices **a** and **b** follows from: $\text{Tr}(\mathbf{ab}) = \mathbf{a} : \mathbf{b}$. Then, by definition, $((\mathbf{a} + \delta\mathbf{a}) : \mathbf{b}) - (\mathbf{a} : \mathbf{b}) = \frac{d(\mathbf{a}:\mathbf{b})}{d\mathbf{a}} : \delta\mathbf{a}$, from which it result $\frac{d(\mathbf{a}:\mathbf{b})}{d\mathbf{a}} = \mathbf{b}^T$.

Thus, from Eq. (33.6) we obtain the expression of the Kalman gain:

$$\mathbf{P}'_k\mathbf{H}^T = \mathbf{K}_k(\mathbf{H}\mathbf{P}'_k\mathbf{H}^T + \mathbf{R}) \rightarrow \mathbf{K}_k = \mathbf{P}'_k\mathbf{H}^T(\mathbf{H}\mathbf{P}'_k\mathbf{H}^T + \mathbf{R})^{-1},$$

that replaced into the expression of $\mathbf{P}_k$ (33.5) leads to

$$\mathbf{P}_k = \mathbf{P}'_k - \mathbf{P}'_k \mathbf{H}^T (\mathbf{H} \mathbf{P}'_k \mathbf{H}^T + \mathbf{R})^{-1} \mathbf{H} \mathbf{P}'_k = \mathbf{P}'_k - \mathbf{K}_k \mathbf{H} \mathbf{P}'_k = (\mathbf{I} - \mathbf{K}_k \mathbf{H}) \mathbf{P}'_k.$$

The projection is achieved from

$$\hat{\mathbf{x}}'_{k+1} = \mathbf{\Phi} \hat{\mathbf{x}}_k,$$

from which

$$\mathbf{e}'_{k+1} = \mathbf{x}_{k+1} - \hat{\mathbf{x}}'_{k+1} = (\mathbf{\Phi} \mathbf{x}_k + \boldsymbol{\omega}_k) - \mathbf{\Phi} \hat{\mathbf{x}}_k = \mathbf{\Phi} \mathbf{e}_k + \boldsymbol{\omega}_k,$$

whose covariance reads

$$\mathbf{P}'_{k+1} = \mathbb{E}[\mathbf{e}'_{k+1} \mathbf{e}'^T_{k+1}] = \mathbb{E}[\mathbf{\Phi} \mathbf{e}_k (\mathbf{\Phi} \mathbf{e}_k)^T] + \mathbb{E}[\boldsymbol{\omega}_k \boldsymbol{\omega}_k^T] = \mathbf{\Phi} \mathbf{P}_k \mathbf{\Phi}^T + \mathbf{Q}.$$

Thus, the Kalman procedure consist of the following steps:

- Kalman gain:
$$\mathbf{K}_k = \mathbf{P}'_k \mathbf{H}^T (\mathbf{H} \mathbf{P}'_k \mathbf{H}^T + \mathbf{R})^{-1};$$

- Update estimate:
$$\hat{\mathbf{x}}_k = \hat{\mathbf{x}}'_k + \mathbf{K}_k (\mathbf{z}_k - \mathbf{H} \hat{\mathbf{x}}'_k);$$

- Update covariance
$$\mathbf{P}_k = (\mathbf{I} - \mathbf{K}_k \mathbf{H}) \mathbf{P}'_k;$$

- Projection
$$\begin{cases} \hat{\mathbf{x}}'_{k+1} = \mathbf{\Phi} \hat{\mathbf{x}}_k \\ \mathbf{P}'_{k+1} = \mathbf{\Phi} \mathbf{P}_k \mathbf{\Phi}^T + \mathbf{Q} \end{cases}.$$

33.5.1 Extended Kalman

The main issue when assimilating data in nonlinear dynamical systems, by employing the so-called extended Kalman filter, concerns the calculation of the Jacobian, whose calculation is computationally expensive when using standard numerical discretization strategies.

In [3] authors proposed a parametric transfer function for data-assimilation in structural dynamics, where at each time step the solution update depends on the model parameters and also on a series of parameters describing the solution at the previous time step. The parametrization of the previous state was performed by invoking the POD and retaining the most significant modes able to approximate any state with the required accuracy.

This parametric solution was then employed for integrating the dynamical system, and its separated representation facilitated the Jacobian calculation, essential in data assimilation in an extended Kalman setting.

33.6 PGD-Based Parametric Inverse Impulse Response, IIR

In structural systems computing the displacements induced by applied forces can be performed very efficiently as described in Chap. 31. However, for control purposes, one would like to infer the forces to be applied in order to ensure a certain displacement. This inverse problem represents a recurrent computational issue. In what follows, first we review the formulation of the direct problem, that is, computing displacements from known forces, by computing and then employing the parametric direct impulse response, DIR. Then, the calculation of the inverse impulse response, IIR, will be discussed.

33.6.1 PGD-Based Direct Impulse Response, DIR

The elastodynamics discrete formulation in the frequency domain, for a loading distributed in space but localized in time, i.e. $\mathbf{F} = \mathbf{f}\delta(t)$, reads

$$(-\omega^2 \mathbf{M} + i\omega \mathbf{C} + \mathbf{K})\mathbf{h} = \mathbf{f}(x),$$

whose PGD-based parametric solution $\mathbf{h}(\omega)$ an be expressed in a separated form

$$\mathbf{h}(\omega) \approx \sum_{i=1}^{N} \mathbf{X}_i \mathcal{W}_i(\omega).$$

When some extra-coordinates (model parameters) are included, here only one, μ, for the sake of simplicity, the parametric direct impulse response reads

$$\mathbf{h}(\omega, \mu) = \sum_{i=1}^{N} \mathbf{X}_i \mathcal{W}_i(\omega) M_i(\mu).$$

By using the inverse transform, its time counterpart can be obtained from

$$\mathbf{h}(t, \mu) = \sum_{i=1}^{N} \mathbf{X}_i \mathcal{F}^{-1}\{\mathcal{W}_i(\omega)\} M_i(\mu) = \sum_{i=1}^{N} \mathbf{X}_i W_i(t) M_i(\mu).$$

33.6.2 Real-Time Monitoring

When the loading exhibits a time evolution given by $P(t)$, the usual procedure consists in computing its Fourier transform, noted by $\mathcal{P}(\omega)$, and then because of the superposition principle, expressing the displacement in the Fourier space as

$$\mathbf{u}(\omega, \mu) = \mathbf{h}(\omega, \mu) \cdot \mathcal{P}(\omega), \qquad (33.7)$$

that constitutes a product of Fourier transforms. Thus, in the time space, Eq. (33.7) results in the convolution product

$$\mathbf{u}(t, \mu) = P(t) * \mathbf{h}(t, \mu) = \int_0^t P(t - \tau) \mathbf{h}(\tau, \mu) d\tau,$$

that can be rewritten in the separated form

$$\mathbf{u}(t, \mu) = \sum_{i=1}^N \mathbf{X}_i \Upsilon_i(t) M_i(\mu),$$

with

$$\Upsilon_i(t) = \int_0^t P(t - \tau) W_i(\tau) d\tau.$$

33.6.3 Inverse Impulse Response

The so-called hybrid laboratories employ experimentation in the part of the system where complex (i.e. not well known) physics happen, and simulation where models can be trusted. Both experiment and simulation communicate via some actuator. Real-time feedback from the simulation is therefore critical. Thus, one would like to compute the forces to be applied to guarantee specified values of the displacement.

For that purpose, the inverse impulse response –IIR– and its parametric counterpart seem compulsory. However, it is well known that direct inversion (at any node $\mathbf{x}_j$ in which the displacement is measured)

$$P(t) = u_j(t) * g_j(t) = \int_0^t u_j(t - \tau) g_j(\tau) d\tau,$$

is ill-posed.

33.6.4 Data-Driven Regularization

One possibility to invert consists in computing by using the DIR the displacement $\mathbf{u}_i(t)$ associated to many loadings $P_i(t)$, $i = 1, \ldots, M$, and then searching $g_j(t)$, $j = 1, \ldots, Q$, by minimizing the regularized mean square error [4]

$$\sum_{i=1}^{M} \|u_{i_j}(t) * g_j(t) - P_i(t)\|^2 + \lambda \|\mathcal{S}(g_j(t))\|^2, \quad j = 1, \ldots, Q,$$

with $\mathcal{S}$ a linear operator.

The same rationale applies in the parametrized case, where both, the displacement and the inverse impulse response are assumed having a separated form [4].

33.7 Inverse Analysis Based on the Reciprocity Principle, RP

33.7.1 Symmetric Operators: The Elastic Problem

We consider the domain Ω, in which an elastic problem is defined, with a load $\mathbf{F}$ applied on the part of the domain boundary Γ_F and the displacements vanishing on Γ_U, with $\Gamma_F \cup \Gamma_U = \Gamma = \partial \Omega$.

When applying the load $\mathbf{F}_1$, the weak form (principle of virtual work) reads

$$\int_\Omega \sigma_1 : \epsilon^* d\mathbf{x} = \int_{\Gamma_F} \mathbf{u}^* \cdot \mathbf{F}_1 d\mathbf{x},$$

that results in the displacement, stress and strain fields, $\mathbf{u}_1$, σ_1 and ϵ_1 respectively.

When applying the load $\mathbf{F}_2$, the weak form

$$\int_\Omega \sigma_2 : \epsilon^* d\mathbf{x} = \int_{\Gamma_F} \mathbf{u}^* \cdot \mathbf{F}_2 d\mathbf{x},$$

results in the displacement, stress and strain fields, $\mathbf{u}_2$, σ_2 and ϵ_2 respectively.

Now, we could consider in the first weak form $\mathbf{u}_2$ and ϵ_2 as test fields, i.e. respectively $\mathbf{u}^*$ and ϵ^*, and $\mathbf{u}_1$ and ϵ_1 as test fields, i.e. respectively $\mathbf{u}^*$ and ϵ^*, in the second one, leading to:

$$\begin{cases} \int_\Omega \sigma_1 : \epsilon_2 d\mathbf{x} = \int_{\Gamma_F} \mathbf{u}_2 \cdot \mathbf{F}_1 d\mathbf{x} \\ \int_\Omega \sigma_2 : \epsilon_1 d\mathbf{x} = \int_{\Gamma_F} \mathbf{u}_1 \cdot \mathbf{F}_2 d\mathbf{x} \end{cases}$$

that in the elastic case, with $\sigma = \mathbf{C} : \epsilon$ (with $\mathbf{C}$ the fourth order elasticity tensor), the left hand members of the previous equation become equal, implying the so-called reciprocity principle

$$\int_{\Gamma_F} \mathbf{u}_1 \cdot \mathbf{F}_2 \, d\mathbf{x} = \int_{\Gamma_F} \mathbf{u}_2 \cdot \mathbf{F}_1 \, d\mathbf{x}.$$

Reciprocity is extremely useful in mechanics. By solving a reference problem numerically (e.g. the second problem), the knowledge of $\mathbf{F}_1$ allows calculating $\mathbf{u}_1$, and vice-versa.

33.7.2 Non-symmetric Operators: The Heat Equation

For deriving the RP in the previous section, the symmetry of the differential operator was crucial. For this reason it does not apply when addressing the heat equation. However, one is tempted to apply the Fourier transform, because by moving to the frequency domain the operator becomes symmetric (even if non-Hermitian).

The price to be paid is operating in the complex space, needing some technicalities, in particular those related to the proper use of the scalar product. Thus, under some conditions [5] the RP continues applying.

References

1. E. Nadal, F. Chinesta, P. Diez, F. Denia, F.J. Fuenmayor, Real time parameter identification and solution reconstruction from experimental data using the proper generalized decomposition. Comput. Methods Appl. Mech. Eng. **296**, 113–128 (2015)
2. R.N. Miller, E.F. Carter, S.T. Blue, Data assimilation into nonlinear stochastic models. Tellus A **51**(2), 167–194 (1999)
3. D. Gonzalez, A. Badias, I. Alfaro, F. Chinesta, E. Cueto, Model order reduction for real-time data assimilation through extended Kalman filters. Comput. Methods Appl. Mech. Eng. **326**, 679–693 (2017)
4. S. Montagud, J.V. Aguado, F. Chinesta, P. Joyot, Parametric inverse impulse response based on reduced order modeling and randomized excitations. Mech. Syst. Signal Process. **135**, 106392 (2020)
5. J.V. Aguado, A. Huerta, F. Chinesta, E. Cueto, Real-time monitoring of thermal processes by reduced order modelling. Int. J. Numer. Methods Eng. **102**(5), 991–1017 (2015)

Open Access This chapter is licensed under the terms of the Creative Commons Attribution-NonCommercial-NoDerivatives 4.0 International License (http://creativecommons.org/licenses/by-nc-nd/4.0/), which permits any noncommercial use, sharing, distribution and reproduction in any medium or format, as long as you give appropriate credit to the original author(s) and the source, provide a link to the Creative Commons license and indicate if you modified the licensed material. You do not have permission under this license to share adapted material derived from this chapter or parts of it.

The images or other third party material in this chapter are included in the chapter's Creative Commons license, unless indicated otherwise in a credit line to the material. If material is not included in the chapter's Creative Commons license and your intended use is not permitted by statutory regulation or exceeds the permitted use, you will need to obtain permission directly from the copyright holder.

Chapter 34
The Twin Continuum

This hybrid approach, where parametric physics-based and parametric data-driven models are combined, is at the origin of the so-called *Hybrid Twin* and hybrid approaches [1–6] revisited in the present section.

34.1 Fully Physics-Based Modelling

Components and systems were usually modeled by using physical principles and laws. Thus, the response of a generic asset to a given action, results from the solution of the problem that involves a physics-based model and the applied loading (the so-called input or action).

In general, models result in a coupled system of linear or nonlinear partial differential equations, that can be solved with the help of powerful computers, when efficiency (accuracy and rapidity) are compulsory, as it is the case in many engineering applications.

As discussed in the present monograph, different techniques were proposed for speeding-up calculations without degrading accuracy, nowadays widely employed in science and technology.

It is worth mentioning that the agreement between the problem solution and the physical asset depends on the pertinence of the considered model to describe the physics involved in its functioning or behavior. When the model describes accurately the analyzed system, simulation becomes the best ally of engineers.

However, such an approach based almost exclusively on the existing physics-based knowledge (in the form of mathematical models), that we will refer as *Virtual Twin*, encounters some major limitations:

1. Its strong dependence on the considered model to finally describe the considered asset. Uncertainty and variability are the main enemies, almost always present in large and complex multiphysics settings..

2. Stability issues, specially noticeable in large systems involving nonlinear dynamics. In these cases predictions and measures can exhibit noticeable bias.
3. The difficulty to cover large systems, finely enough for resolving all the space and time scales.
4. The computational cost for solving the associated problems, with the required accuracy, in a time compatible to its use for control purposes, needed in the nowadays engineering applications, more concerned by performances in operation than by the design itself.

It is important to mention that nowadays engineering is expected accompanying each asset all along its life, whose diagnosis, prognosis and the associated prescriptive actions and decision making, must be compatible with the asset characteristic time. Thus, new industry (industry 4.0) and society (smart city, smart nation, ...) need more stringent performances that are not compatible with usual simulation-based engineering procedures.

Model order reduction, surrogates, meta-models, ... largely treated in the present monograph, allowed to address partially the main just referred issues, in particular the last, related to computational efficiency. It is also important to mention that, model order reduction is in many cases (in particular when using surrogates or meta-models) based on the use of advanced machine learning techniques revisited before, but operating on synthetic data coming from fine-grained models solved with the appropriate discretization techniques. Even if that data production is in general extremely expensive (computationally speaking), it operates offline.

34.2 Fully Data-Driven Modeling

To circumvent the just referred limitations of the well stablished physics-based modeling paradigm, in particular: (i) the ones encountered when addressing systems poorly known, whose mathematical models do not exist, or they remain too poor for predictive purposes; or (ii) when real-time feedbacks become compulsory, challenging even the most powerful and efficient reduced order models (because of stability issues, multi-scale complexity, the problem size or its nonlinearity, to cite few), data-driven modeling irrupted and rapidly developed to conquer numerous science and engineering disciplines. These data-driven models are nowadays highly present in many technologies concerning autonomous systems, enabling real-time responses, while ensuring the required accuracy.

This impressive growing was impulsed by the advances and democratization of measurement devices, data-collection and communication (IoT), data storage and manipulation at different levels (edge, cloud, ...) and new computer architectures able to make fast and cheap (the energy cost is a term in the equation).

However, things were, and are being, a bit more delicate than initially expected. In particular:

- In many technological applications, data is not so abundant as wished. Data is expensive and difficult to acquire, and even sometimes its acquisition is forbidden by regulations.
- The accuracy of measurement devices must be integrated into the learnt models, and limits the scales attainable by the knowledge.
- The cost of performing measurements in engineering (because of both, the devices and the measurement process), implies the need of replacing the big-data paradigm by another more valuable, involving the right data at the right place and time, a sort of smart-data paradigm,
- Models and the derived decisions, should be explained when addressing critical systems that must be certified before being used. Explainable AI is a timely topic of major interest. Validation and verification must also to be urgently adapted to data-driven modeling frameworks.
- Finally, extrapolation, remains an important issue of data-driven models, fact that motivates the interest of discovering physics more than simply inferring the state.

The virtual replica of an asset fully-based on the use of data, is here called *Digital Twin*.

34.3 Just in Between Both Paradigms

The just referred learning paradigms, physics-based and data-driven, offer many, major and unquestionable benefits. However, one question remains open: there is a twining continuum, in between the two extrema, the fully physics-based (virtual twin) and the fully data-driven (digital twin) that could ally the advantages of both, while circumventing the main limitations of both?

For that purpose two recent frameworks appeared and are conquering many domains of science and technology.

34.3.1 Physics-Aware Learning

Physics-aware learning tries to incorporate knowledge, physics or first principles, during the learning process.

PINN (Physics Informed Neural Networks), previously addressed in the present monograph, approximates the problem solution by using NN, and reduced the amount of data for training it, while increasing the explicability of the computed solution, by enforcing NN to satisfy the partial differential equation, PDE.

The fact of employing the PDE, empowers the learning process, that now needs much less data. The regression is constructed by enforcing the vanishment of the PDE residual, instead of using collected data. Data, is here used for mitigating missing information, for example concerning loading or the boundary conditions.

Sometimes, to reduce such a strong formalism that assumes that the PDE performs well with the addressed system, only first principles are enforced, as structure preserving NN considers, where only first principles, e.g. energy conservation and entropy production are enforced.

Finally, by using adequate techniques, domain knowledge can be assimilated during the learning process, by addressing their associated constraints in a weak or strong form [7, 8], and some constraints can be enforced hardly into the neural networks architecture, as for instance convexity [9].

34.3.2 Physics-Augmented Learning

Another valuable approach consists of assuming that the reality can be expressed by a physics-based model, that constitutes a valuable approximation, complemented by a data-driven model of the discrepancy, the so-called ignorance.

Hybridation can follow two different routes, the one enriching the solution and the one enriching the model, from which the enriched solution will result.

34.3.2.1 Solution Enrichment

With the physics-based model given operating in real-time by making use of model order reduction techniques previously introduced, and with the data-driven correction of the solution learnt by using an appropriate regression technique (among the numerous possibilities previously introduced) on the deviation data, the expected solution consists of the addition of the physics-based and the data-driven ones [3, 4, 10].

Thus, online, as soon as data is collected, the model is calibrated, that is, the model parameters identified, by trying to fit the physics-based model prediction to the measurements. Then, the data-driven model is particularized for the just identified values of the parameters, and then the hybrid prediction obtained.

Other variants exist, where calibration and the data-driven model training are performed simultaneously, particularly important in the case of dynamical systems, where both models should remain coupled when integrating in time. A decoupled formulation, much less performant, is considered when the physics-based model is a black-box, making impossible its coupling with the data-driven model at each time step.

34.3.2.2 Model Entichement

We consider the nominal model given by the matrix form $\mathbf{MU} = \mathbf{F}$, where $\mathbf{F}$ and $\mathbf{U}$ are, when modeling a simple elastic problem, the nodal force and displacement vectors, respectively.

Now, displacement is collected at some locations and grouped into the vector $\mathbf{U}^{exp}$. At these locations, the nominal model predicts the displacements grouped into the vector $\mathbf{U}^{pred}$, and a noticeable gap is noticed, that is, the norm of the displacements difference $\|\mathbf{U}^{exp} - \mathbf{U}^{pred}\|$ remains greater than the assumable prediction error.

Thus, it seems that the model $\mathbf{M}$ remains unable to represent the measurements $\mathbf{U}^{exp}$, and then, it needs an enrichment $\Delta\mathbf{M}$, such that the enriched model $\mathbf{M} + \Delta\mathbf{M}$ should represent the collected data, that is, the corrected model $\mathbf{M} + \Delta\mathbf{M}$ and the associated displacement prediction $\mathbf{U} + \Delta\mathbf{U}$, should verify the equilibrium, i.e. $(\mathbf{M} + \Delta\mathbf{M})(\mathbf{U} + \Delta\mathbf{U}) = \mathbf{F}$ while representing the collected data, i.e. $\|(\mathbf{U} + \Delta U)^{pred} - \mathbf{U}^{exp}\|$ remains small enough.

Now, a parametrization of the model correction combined with an adequate regularization could serve to compute the model enrichment $\Delta\mathbf{M}$ while performing the data completion $\Delta\mathbf{U}$.

The just referred problem can be formulated in a mathematical setting [11] or in physics-augmented neural network setting [12].

When following this route, the main issues concern: (i) the choice of the functional to be minimized for performing the hybridation; (ii) the enrichment parametrization (reduced basis, matrix parametrization, ...); and (iii) ensuring the properties on the enriched model (e.g. stability when addressing transient simulations, positivity, ...), the last deeply considered in [5, 6].

References

1. R. Ibanez, E. Abisset-Chavanne, D. Gonzalez, J.L. Duval, E. Cueto, F. Chinesta, Hybrid constitutive modeling: data-driven learning of corrections to plasticity models. Int. J. Mater. Form. **12**, 717–725 (2019)
2. D. Gonzalez, F. Chinesta, E. Cueto, Learning corrections for hyper-elastic models from data. Front. Mater.-Sect. Comput. Mater. Sci. (2019). https://doi.org/10.3389/fmats.2019.00014
3. F. Chinesta, E. Cueto, E. Abisset-Chavanne, J.L. Duval, F. El Khaldi, Virtual, digital and hybrid twins: a new paradigm in data-based engineering and engineered data. Arch. Comput. Methods Eng. **27**, 105–134 (2020)
4. B. Moya, A. Badias, I. Alfaro, F. Chinesta, E. Cueto, Digital twins that learn and correct themselves. Int. J. Numer. Methods Eng. **123**(13), 3034–3044 (2022)
5. A. Sancarlos, M. Cameron, A. Abel, E. Cueto, J.L. Duval, F. Chinesta, From ROM of electrochemistry to AI-based battery digital and hybrid twin. Arch. Comput. Methods Eng. **28**, 979–1015 (2021)
6. A. Sancarlos, M. Cameron, J.M. Le Peuvedic, J. Groulier, J.L. Duval, E. Cueto, F. Chinesta, Learning stable reduced-order models for hybrid twins. Data Centric Eng. **2**, e10 (2021)
7. T. Beucler, M. Pritchard, S. Rasp, J. Ott, P. Baldi, P. Gentine, Enforcing analytic constraints in neural networks emulating physical systems. Phys. Rev. Lett. **126**(9), 098302 (2021)
8. P. Marquez-Neila, M. Salzmann, P. Fua, Imposing Hard Constraints on Deep Networks: Promises and Limitations (2017). arXiv:1706.02025v1
9. B. Amos, L. Xu, J.Z. Kolter, Input convex neural networks, in international conference on machine learning (PMLR, 2017), pp. 146–155
10. V. Champaney, F. Chinesta, E. Cueto, Engineering empowered by physics-based and data-driven hybrid models. Int. J. Mater. Form. **15**, 31 (2022)

11. D. Di Lorenzo, V. Champaney, C. Germoso, E. Cueto, F. Chinesta, Data completion, model correction and enrichment based on sparse identification and data assimilation. Appl. Sci. **12**, 7458 (2022)
12. D. Di Lorenzo, V. Champaney, J.Y. Marzin, C. Farhat, F. Chinesta, Physics informed and data-based augmented learning in structural health diagnosis. Comput. Methods Appl. Mech. Eng. **414**, 116186 (2023)

Open Access This chapter is licensed under the terms of the Creative Commons Attribution-NonCommercial-NoDerivatives 4.0 International License (http://creativecommons.org/licenses/by-nc-nd/4.0/), which permits any noncommercial use, sharing, distribution and reproduction in any medium or format, as long as you give appropriate credit to the original author(s) and the source, provide a link to the Creative Commons license and indicate if you modified the licensed material. You do not have permission under this license to share adapted material derived from this chapter or parts of it.

The images or other third party material in this chapter are included in the chapter's Creative Commons license, unless indicated otherwise in a credit line to the material. If material is not included in the chapter's Creative Commons license and your intended use is not permitted by statutory regulation or exceeds the permitted use, you will need to obtain permission directly from the copyright holder.

Chapter 35
Conclusion

The present review summarized a diversity of technologies in different domains, the one of data, the one related to the efficient solution of physics-based models, the one of learning models from data, and the ones of knowledge and data hybridation and data assimilation, major protagonists of an incipient new engineering, allying efficiency and accuracy, based on the employ of the so-called physics informed digital twins.

The present review aims at giving some fundamentals and valuable references to make possible deeper studies.

Open Access This chapter is licensed under the terms of the Creative Commons Attribution-NonCommercial-NoDerivatives 4.0 International License (http://creativecommons.org/licenses/by-nc-nd/4.0/), which permits any noncommercial use, sharing, distribution and reproduction in any medium or format, as long as you give appropriate credit to the original author(s) and the source, provide a link to the Creative Commons license and indicate if you modified the licensed material. You do not have permission under this license to share adapted material derived from this chapter or parts of it.

The images or other third party material in this chapter are included in the chapter's Creative Commons license, unless indicated otherwise in a credit line to the material. If material is not included in the chapter's Creative Commons license and your intended use is not permitted by statutory regulation or exceeds the permitted use, you will need to obtain permission directly from the copyright holder.

The manufacturer's authorised representative in the EU is Springer Nature Customer Service Centre GmbH, Europaplatz 3, 69115 Heidelberg, Germany. If you have any concerns regarding our products, please contact ProductSafety@springernature.com

Printed and bound by CPI Group (UK) Ltd, Croydon, CR0 4YY
26/03/2026
02078941-0003